Evan R. Ward

Hubbing for Tourists

Evan R. Ward

Hubbing for Tourists

Airports, Hotels and Tourism Development
in the Indo-Pacific, 1934–2019

DE GRUYTER
OLDENBOURG

ISBN 978-3-11-132486-9
e-ISBN (PDF) 978-3-11-132664-1
e-ISBN (EPUB) 978-3-11-132763-1

Library of Congress Control Number: 2023940480

Bibliographic information published by the Deutsche Nationalbibliothek
The Deutsche Nationalbibliothek lists this publication in the Deutsche Nationalbibliografie;
detailed bibliographic data are available on the internet at http://dnb.dnb.de.

© 2023 Walter de Gruyter GmbH, Berlin/Boston
Cover image: Matteo Morando, https://en.wikipedia.org/wiki/File:JewelSingaporeVortex1.jpg,
no changes were made to the image
Typesetting: Integra Software Services Pvt. Ltd.
Printing and binding: CPI books GmbH, Leck

www.degruyter.com

Acknowledgements

I am grateful to God for guidance in finding the papers I needed to write this book. I am also grateful for my parents. From a young age, my father took me to airport terminals, which were of equal fascination to the planes we spotted.

A number of people helped with the funding of this project. At the University of North Alabama, my former chair, Dr. Daniel Heimmermann, made funds available to purchase the initial tranche of papers from the Frank Lloyd Wright Archives for the first chapter of the book. At Brigham Young University, the Kennedy Center provided a travel grant to a conference in Ghana, which allowed me to travel to Dubai for the first time, an experience that sparked the writing of this book using the papers provided by Dr. Heimmermann.

Also at Brigham Young University, my former chair, Dr. Eric Dursteler, made funding available for my research trip to Singapore. My current chair, Dr. Brian Q. Cannon, made funding available to do research in Australia and New Zealand, as well as support for a conference at Cambridge University, which allowed me to visit the National Archives at Kew Gardens and the Special Collections at the School of Oriental and African Studies at the University of London.

Staff at several archives, as well as an anonymous fellow researcher, provided invaluable support for the project. I am grateful for the staff at the National Archives in Kew Gardens, the National Archives of Singapore, and National Archives of Australia (at both the Canberra and Melbourne locations). I cannot thank enough the young fellow researcher in Canberra who tipped me off to the work of former Qantas CEO, John Menadue, who charted an intra-Asian strategy for the national flag-carrier. I was subsequently able to contact Mr. Menadue, who was gracious enough to grant me a telephone interview. At the National Library of Australia (Canberra), I encountered another miracle. I did not know about Mr. Menadue's work until I received the tip at the National Archives of Australia. The young researcher told me that his papers were at the National Library across town. To my chagrin, I learned that the papers had not been processed and that such a procedure would take weeks at best. Lo and behold, when I returned two days later to consult other papers at the Library's Special Collections, the cache of hundreds of pages of speeches was ready for my perusal. Imagine, there was *one* archival box in the whole of Australia that held the key to my research, and I was able to find it, but only with significant help.

At the School of Oriental and African Studies, University of London, Swire Archivist, Julie Makison, provided access to speeches delivered by Swire executives. Her organizational skills made my surgical visit to the University of London Archives swift and successful.

https://doi.org/10.1515/9783111326641-202

Although this study began with a focus on Dubai and Singapore, it would not have been complete without the exceptional help I received in Hong Kong. Staff at the Government Records Office quickly found documents relevant to my project. Matthew Edmundson, archivist at the Swire Group, parent company of Cathay Pacific Airways and low-cost carrier, Hong Kong Express, selected company newsletters relevant to my manuscript. Finally, Ms. Natalie Kong, Communications Coordinator made significant efforts on my behalf to arrange for a tour of the Peninsula Hotel, as well as put me in touch with the Melanie Li at the Hong Kong Heritage Project, a visit which yielded significant archival finds related to aviation and tourism in Hong Kong. Ms. Carrie Chen and Ms. Carolyn Lee patiently answered queries about The Peninsula and its history as an incomparable stop-over hotel. Ms. Adrienne Tin facilitated my visit as well. Finally, at the Hong Kong Heritage Project, Edith carefully guided me through the materials there.

Bertram Gordon, Shelley Baranowski, and Eric Zuelow offered helpful advice in navigating the selection of an appropriate press. I am deeply indebted to the reviewers of two articles published in *The Journal of Tourism History* for their comments, which opened the door to an Australian angle to this book. Thoughtful reviewers from DeGruyter Press were equally as helpful, recommending that I add chapters on Dubai and another on low-cost carriers in preparing the final book. Taylor and Francis granted permission to use two articles from the Journal of Tourism History for the book. These are "Before Dubai: the Frank Lloyd Wright Foundation and Iranian tourism development, 1967–1969," volume 11, number 1, pages 46–62 (https://www.tandfonline.com/doi/abs/10.1080/1755182X.2019.1567830); and "'An Entrepot for Tourism': From Developmental Tourism to Gateway Travel in Singapore, 1937–2019," volume 14, number 1, 1–27 (https://www.tandfonline.com/doi/abs/10.1080/1755182X.2022.2091667?journalCode=rjth20).

I am most grateful to Ms. Rabea Rittgerodt, the Senior History Acquisitions Editor at DeGruyter Press, for making the publication of this book such a delightful process. It has been a thrill to write about what fascinates me most in life.

Lenore Carrier and Jen Nelson provided administrative support for purchasing materials. A number of front desk associates, including Macey Chatterton, printed countless copies of my revisions.

My Aunt and Uncle, Amy and Wayne Willis, provided a place to stay, hospitality, and logistical assistance during my visit to Hong Kong.

Finally, my wife, Jennie, and children, Sydney and Sam, who have accompanied me on some journeys, provided encouragement throughout the research and writing process. There's no better place than home.

I conclude noting that I finished this manuscript on April 28, 2023, aboard Korean Airlines flight 172 between Hong Kong's airy Chek Lap Kok International Airport and Asia's supreme temple to aviation, Seoul's immaculate Incheon International Airport.

Note: Except in the context of historical quotes, I have employed current place names for cities and nations referenced in the book. This is for the sake of consistency, even where names and nations have changed borders and titles over time. In referencing individuals, I have maintained traditional practices for the countries of origin (e.g., using the surname first when discussing individuals of Chinese descent) unless that individual has chosen to publicly be referenced according to Western conventions (i.e., given name first then surname). Surname primacy has been maintained in the index except in cases where they are not used, such as with some Malay names. I, of course, am responsible for the content, but grateful for the abundance of help offered while writing the book.

Evan R. Ward
Brigham Young University

Contents

Introduction

Hubbing for Tourists: Airport, Hotel and Tourism Development in the Indo-Pacific, 1934–2019

The sense of expectation was palpable as the giant Airbus A380-800 came into view beneath the clouds and headed towards the Burj Al-Arab [Hotel]. When the superjumbo swooped low near the hotel, the crowds lining Jumeirah Beach gasped with amazement. The aircraft glided serenely past the Burj, then twisted quickly left and right before turning to start another approach. The Emirates-liveried A380 made about a dozen passes near the seven-star Burj Al-Arab.
– *Gulf News* at unveiling of Emirates' A380 Fleet[1]

Plans [are] going ahead for the official opening of the new Gulf Hotel in Bahrain . . . The hotel [will] be run by [Gulf Air] . . . British Overseas Aviation Corporation proposed to tell Shaikh Rashid in London that they would be interested in a similar project in Dubai; plans were going ahead for new hotels under similar arrangements in Abu Dhabi and Doha.
– British communique, 1969[2]

A battle is under way for airspace in Asia these days – and it has nothing to do with American spy aircraft off China. It has to do with bureaucrats, airlines and construction firms vying for their cities to become – or in some cases remain – hub airports for their particular areas.
– *Far Eastern Economic Review*, 2001[3]

The airport layover is often dismissed as inconvenient – but with a little planning, it can add significantly to the holiday experience. Particularly if you interrupt your journey to enjoy an extended spell – a day, even two – in the city where you will be changing flight . . . You get to explore a second destination, probably on a different continent.
– Chris Leadbetter, "Kickstart your Holiday with a Stopover"[4]

Dubai International Airport (DXB), Emirates Airlines, and the Burj al-Arab. Changi International Airport (SIN), Singapore Airlines, and Marina Bay Sands Hotel. Chek Lap Kok (HGK), Cathay Pacific, and The Peninsula Hotel. Kingsford Smith (SYD),

---

1 *Gulf News* as quoted in Graeme Wilson, *Emirates: The Airline of the Future* (Dubai: Prima Media, 2007), 327.
2 "Call by Mr. Orpin of BOAC," in "Visit of the Ruler of Dubai and Aviation Law in the Persian Gulf," FCO 14/591, Arabian Gulf Digital Archives., https://www.agda.ae/en (cited hereafter as "AGDA"). Documents from the AGDA originate at the British National Archives.
3 John Larkin, "Focus: Asian Infrastructure: Airports; Flying High – In the Race to Capture Asia's Growing Travel Market, Cities are Competing to Build Airports that are Bigger, Better and Cheaper," *Far Eastern Economic Review*, May 10, 2001, Factiva.
4 Chris Leadbeater, "Kickstart Your Holiday with a Stopover," *The Daily Telegraph*, October 22, 2022, Factiva.

https://doi.org/10.1515/9783111326641-001

Qantas Airlines, and the Wentworth Hotel. What do these collective entities have in common? Not only do they link global air hubs with city-centric long-haul airlines and destination-worthy hotels, but they are the product of a distinct strategy to boost tourism development through the synergies created by aviation development. The regional genesis of such a strategy took advantage of (1) the airport infrastructure boom throughout the Indo-Pacific during the late twentieth and early twenty-first century;[5] (2) the success of long-haul airlines supported by ambitious city-states and propelled by innovation in wide-body airplanes; and (3) the economies of scale created by iconic hotel properties tightly aligned with their city's global aspirations for preeminence among jet-setting tourists. As Asian aviation specialists Kevin O'Connor and Kurt Fuelhart have observed, "Put simply, places that are developed as tourist destinations attract air traffic, while air traffic patterns can facilitate tourism development. These interdependencies have a special relevance in the Asia Pacific region, as many nations have integrated tourism infrastructure and airline activity as part of broader economic development strategies."[6]

This book argues that tourism development in Indo-Pacific city-states followed a logic distinct from that proposed by tourism consultants, and multilateral aid organizations (including UNESCO, the World Bank, and The Organization of American States) in other emerging parts of the world, including Latin America and Africa. While Dubai, Singapore, Hong Kong, and Shanghai did not eschew international aid entirely in tourism development, they principally nurtured global ambitions that drew on export-led solutions sustained by state-led initiatives (in airlines and globally competitive air hubs) and private investments (hotel development).

Although these stories played out against the backdrop of decolonisation, the conditions for their rise as aviation entrepots predated independence. Most importantly, the history of each of these city-states was colonial in nature. As a result, they enjoyed global linkages to cities beyond their hinterlands, a circumstance that facilitated global gateway status.[7]

---

5 As early as 1993, journalist Jahabar Sadiq noted that "at least 10 major new airports are now on the drawing board or under construction, with many more being expanded or undergoing extensive facelifts to cope with the predicted growth of air travel in Asia." See Sadiq, "Sepang Airport Seen Facing Competition," *Business Times*, June 14, 1993, Factiva.
6 Kevin O'Connor and Kurt Fuelhart, "The Asia Pacific Region and Australian Aviation," in David Timothy Duval, ed., *Air Transport in the Asia Pacific* (Farnham, UK: Ashgate, 2014), 76.
7 Timothy Vowles and Daniel Mertens write, "It is . . . believed that having a recent colonial past aid in the emergence of primate cities . . . This historical circumstance has enabled them to create strong linkages to former colonial powers, which has aided their becoming global gateways,"

The status of Dubai, Singapore, Hong Kong, and Shanghai as stopovers, both in the age of steamships and with the opening of the Suez Canal, as well as in the age of commercial aviation, also preceded decolonisation. Dubai established its reputation as a trading port long before Emirates Airlines took flight. An Imperial Airways rest house attached to nearby Sharjah airfield preceded the later construction of the "seven-star" Burj al-Arab Hotel. Singapore boasted the incomparable Raffles Hotel; owned, curated, and managed by the Armenian Sarkies family at the turn of the twentieth century. This distinction only slowly ceded market share to overseas-Chinese hoteliers as shopping centers blossomed on Orchard Road in the early 1970s. Hong Kong reigned as Asia's hotel trendsetter, beginning with the Hongkong Hotel in the late nineteenth century, only to be surpassed by the Kadoorie family's Repulse Bay Resort in 1920 and the legendary Hong Kong Peninsula in 1928. These early palisades of pleasure held their own, even as the Mandarin Oriental (of Jardine Matheson fame) launched its own international diaspora in the 1960s and 1970s. Finally, no single property in Asia enjoyed the splash and over the top, sixteen story decadence like Victor Sassoon's garish Cathay Hotel, opened on Shanghai's Nanjing Road at the Bund.

What *was* pre-dominantly post-colonial was the emergence of city-state-based airlines linking the rest of their respective hinterlands to the wider world. Curious tourists, who often encountered these cities first on stopovers at gleaming air terminals, returned to the entrepots as destinations themselves. Cathay Pacific would be decidedly colonial, not only because the Swire Company initiated scheduled flights within the region soon after World War II, but also because Hong Kong remained a British outpost until 1997. In the case of Singapore, its hometown airline devolved from Malayan-Singapore Airlines in the late 1960s. The new city state took possession of the long-range Boeing 707s and existing routes to international destinations. Farther west, the Al-Maktoum family launched Emirates in 1985 as a challenge to Gulf Air. Emirates leased its first planes from Pakistan, signaling ambitions far beyond Bahrain, Oman, and neighboring emirates. Finally, the genesis of China Eastern Airlines as Shanghai's "hometown" carrier gained credence with the construction of Pudong International Airport in 1999. Challenging Hong Kong, the new airport vied for the title of China's primary gateway to the mainland.

If a study about hubs, this book also examines the peripheries it connected. Due to a flood of sources made available by the various Australian local, state, and national governmental organisations' digitization of periodicals and newspapers, we learn a lot about the economic, social, and cultural motivations of those

---

in "Gateway Airports and International and Regional Connectivity of Air Transport in the Asia Pacific," in Duval, ed., *Air Transport in the Asia Pacific*, 116.

individuals traveling through these hubs. Their antipodal circumstances primed them to travel, beginning with commercial flights on Empire Airways to London through Project Sunrise, set to link London and Sydney directly in the mid-2020s. At the same time, Australia's Asian orientation continues to evolve, making it an important part of the story of entrepot-based tourism in the twenty-first century.[8]

Intellectually, this book began as a study of why and how the Persian Gulf developed regional entrepot facilities for entertainment, shopping, and transit-centric aviation. While the story of Dubai's ascendence is relatively well known, other efforts to create tourist-centric hubs materialised before the rise of the glittering emirate on the Persian Gulf. In the late 1960s, Iran's Shah Reza Pahlavi commissioned Frank Lloyd Wright's Taliesin West architectural studio to conceptualise an exotic tourist destination on the island of Kish, located off Iran's southwestern coast. The project at Kish included lavish hotels and attractions (including a Persian-themed amusement park) that appealed to a predominantly Arab clientele. Beirut had famously served as their former stomping grounds. Part of the initial project included a massive aviation entrepot that would situate Iran at the crossroads between Europe and Asia, a clear forerunner of the modern-day Persian Gulf airports at Doha, Dubai, and Abu Dhabi. Under such a scheme, Iran Air was the chosen instrument to link the far-flung continents. Although the scale of the project never matched the Shah's vision, it represented an attempt to establish alternative networks of commercial aviation alongside those already established by Western airports and airlines in the 1960s. Chapter 1 draws on the work of Waleed Hazbun and Arang Keshavarzian, each of whom articulated the idea that Middle Eastern governments and actors worked to control the production of tourism and transport to regions often overlooked by their Western counterparts.[9] Chapter 1, a version of which appeared in the *Journal of Tourism History*, sets out the model for indigenous regional aviation and tourism development that appealed particularly to an Arab clientele.[10] While these efforts failed to take root in Iran, they caught on in Dubai and elsewhere throughout the Gulf states. I have refrained from telling

---

8 See O'Connor and Fuelhart, who write, "So today, Singapore, Thai, Cathay, Malaysian, JAL, Garuda, Korean Airlines, and China Eastern aircraft are seen at Australian airports, providing services to cities such as Beijing, Shanghai, Guangzhou, Ho Chi Minh City, Seoul and [Bali] – all of which played little or no part during earlier periods" (74).

9 Waleed Hazbun, *Beaches, Ruins, Resorts: The Politics of Tourism in the Arab World* (Minneapolis: University of Minnesota Press, 2008); Arang Keshavarzian, "Geopolitics and the Genealogy of Free Trade Zones in the Persian Gulf," *Geopolitics* 15 (2010), 263–289.

10 See Evan R. Ward, "Before Dubai: The Frank Lloyd Wright Foundation and Iranian Tourism Development, 1967–1969," *Journal of Tourism History* 11 (2019), https://www.tandfonline.com/doi/full/10.1080/1755182X.2019.1567830?src=recsys, accessed January 14, 2023.

the entire story of Dubai's rise yet again, given the superlative work of other journalists, scholars, and public historians.[11] I conclude chapter 1 with the early story of Dubai insofar as I am able to use primary sources from the British National Archives to illuminate the transformative impact of Dubai's rise as an Indo-Pacific commercial aviation entrepot. In the footnotes, I include portions of an oral history I conducted via telephone with architect Nezam Amery years ago.

The second chapter explores how Singapore developed as a tourism-centric aviation entrepot. I apply Hazbun's and Keshavarzian's models across regions to emphasise why the Lion City pioneered its own long-haul airline, developed a hub-

---

11 Among historians, Graeme H. Wilson's unparalleled access to archives and individuals associated with the rise of Dubai, its commercial aviation sector, and the special role of shopping at airports deserves mention, if not wider distribution given the difficulty of accessing archival materials regarding institutions such as Emirates Airlines. For a broad overview of the rise of Dubai, see Wilson, *Rashid's Legacy: The Genesis of the Maktoum Family and the History of Dubai* (London: Media Prima, 2006). Wilson enjoyed broad access to the prime movers in the development of Emirates Airlines, the results of which are published in *Emirates: The Airline of the Future* (Media Prima, 2007). Wilson treats the broad sweep of commercial aviation in the emirate in *Flight into the Future: Seventy Years of Civil Aviation in Dubai* (London: Media Prima, 2007). Finally, Wilson explores the niche that set Dubai apart among its hubbing counterparts in the Indo-Pacific in *Fly Buy Dubai: The Remarkable 25 Year Journey of Dubai Duty Free* (London: Media Prima, 2008). John D. Kasarda and George Lindsay chronicle the evolution of Dubai's airport and the rise of Emirates Airlines in their conceptually pathbreaking work, *Aerotropolis: How We'll Live Next* (New York City: Farrar, Straus, and Giroux, 2011), passim. Journalists have also recounted the surprising rise of the city-state, including the role played by commercial aviation and tourism. See Jim Krane, *City of Gold: Dubai and the Dream of Capitalism* (New York City: St. Martin's Press, 2009). Daniel Brook offers a comparative look at Dubai vis-à-vis other brash commercial upstarts including St. Petersburg, Shanghai, and Mumbai in *A History of Future Cities* (New York City: W.W. Norton, 2013). In a more scholarly vein, Christopher Davidson explores the inter-locking governmental and commercial structures of the emirate in *Dubai: The Vulnerability of Success* (New York City: Oxford University Press, 2008). Neha Vora examines the complexities of city-state citizenship in *Impossible Citizens: Dubai's Indian Diaspora* (Durham, North Carolina: Duke University Press, 2013). Todd Reisz's important study, *Showpiece City: How Architecture Made Dubai* (Palo Alto, California: Stanford University Press, 2020), examines the British influence on large scale building projects in the emirate during its' transition from Trucial colony to independent emirate. Although he does not treat aviation infrastructure, his book is invaluable for demonstrating the sharing of knowledge for later, more ambitious projects that drew on Western expertise. Amelie Le Renard's recent volume (with Jane Kuntz as translator) explores the social and cultural landscapes of the twenty-first century experience through the eyes of Europeans, among others, in *Western Privilege: Work, Intimacy, and Post-Colonial Hierarchies in Dubai* (Palo Alto, California: Stanford University Press, 2021). More broadly, Rory Miller examines the rise of Dubai against the broader phenomenon of the emergence of the Gulf city-states in *Desert Kingdoms to Global Powers: The Rise of the Arab Gulf* (New Haven, Connecticut: Yale University Press, 2016).

bing strategy catering to tourists visiting other parts of Southeast Asia, and ultimately, transformed their Changi International Airport into a tourist destination itself. The chapter also draws on studies charting intra-Asian networks past and present. Key among these works include the research of Engseng Ho, who identifies historical intra-Asian networks renewed in the contemporary era.[12] This aligns with the scholarship of Sugata Bose, whose work on the greater Indian Ocean Basin illustrates that the new soundings throughout the basin, be they in shipping, trade, or tourism, had roots in a largely inter-dependent, non-hegemonic system that was only thrown into chaos by European colonialism.[13] My study suggests that the new aviation networks established in Singapore, as well as at competing city-states, including Hong Kong, Dubai, Shanghai, and Mumbai, fit among these reified systems in a dynamic region integrated between regional and global participants.

Chapter 3 applies the argument of indigenous hotel development schemes as part of the "hubbing for tourists" process in the Asia Pacific generally, and Southeast Asia particularly. As such, it highlights how those institutions emerged against the backdrop of commercial aviation in the broader Indo-Pacific context. While the preponderance of previous work on Asian hotels deals with questions of colonialism, architecture, and the commercialisation of heritage properties, I integrate Gordon Pirie's foundational study on the emergence of "incidental tourists" on the air routes linking Asia, Europe, and Australia, with Maurizio Peleggi's scholarship on the grand hotels of Southeast Asia strung along the steamship routes between Aden and Hong Kong to demonstrate how hotel networks in Asia developed in tandem with air services. They then evolved, as the important design work of Eunice Seng emphasises in her study of the modern Southeast Asian hotel, into hotel chains created by overseas-Chinese, enterprising Indians, and Baghdadi hoteliers in Hong Kong.[14] Quite simply, as widebody jets disgorged exponentially larger groups of tourists in Asia, hotels were needed to lodge them. Structurally, the chapter unfolds with an assessment of how hotels and commercial aviation developed hand-in-hand prior to World War II; then, examines the transition to Asian-based hotel chains that flourished side-by-side with American

---

12 Engseng Ho, "Inter-Asian Concepts for Mobile Societies," *The Journal of Asian Studies*, 76: 4 (November 2017), 907–928.
13 Sugata Bose, *A Hundred Horizons: The Indian Ocean in the Age of Global Empire* (Cambridge, Massachusetts: Harvard University Press, 2006).
14 See Gordon Pirie, "Incidental Tourism: British Imperial Air Travel in the 1930s," *Journal of Tourism History*, 1:1 (2009), 49–66; Maurizio Peleggi, "The Social and Material Life of Colonial Hotels: Comfort Zones as Contact Zones in British Colombo and Singapore, ca. 1870–1930," *Journal of Social History*, 46:1 (Fall 2012), 124–153; Eunice Seng, "Temporary Domesticities: The Southeast Asian hotel as (Re)presentation of Modernity, 1968–1973," *The Journal of Architecture*, 22:6 (2017), 1092–1136.

chains after the advent of widebody travel; and, finally, briefly lays out the culmination of the relationship between hubs and hotels with the emergence of the latter as attractions unto themselves in Dubai and Singapore.

Chapter 4 presents a model of passengers and tourists passing through the Australia to Europe corridor, often known as the "Kangaroo Route." Given Qantas Airlines' development of this corridor, the Australian inclination for long-haul travel, and the abundance of archival sources, Australians are the social focus for this chapter. For a country whose signature history of identity is entitled *The Tyranny of Distance: How Distance Shaped Australia's History*, long-haul aviation was second nature in the twentieth and twenty-first century. Beginning with London to Australia flight service in 1934, Australian public figures, Qantas executives, and a mass of tourists, critiqued airports and airlines as keenly as any other national collective during the modern era. This aspect of Australian life also offered a measuring rod against which integration into the Asia-Pacific could be compared. What once had been an arms-length relationship, which Qantas executives and the flying public simply hoped to fly over, became part of an integrated network clearly entrenched in the region by the turn of the twenty-first century. With Qantas and budget-minded subsidiary Jetstar, Australia was no longer a "bridge" to the Asia-Pacific; arguably, it was as much Asian and Pacific as its neighbors in what former Prime Minister Robert Menzies denominated "the near north."

Chapter 5 deals with the potentially disruptive threat of low-cost-carriers to the "hubbing for tourists" model. Urban planning scholar Max Hirsh makes the case that the newest travelers in Southeast Asia demand an infrastructure for flight adapted to their more modest means.[15] Using the latest (cited hereafter as "AGDA") scholarship on emerging trends in low-cost-carrier terminals, as well as cutting edge analysis from scholars of low-cost-carriers in the emerging world, including John Bowen, Shelley Vishwajeet, Sven Gross and Michael Lück, I argue that low-cost-carriers and airport administrators have *accelerated* hubbing with leading edge technology meant to reduce the cost of airport services to LCC's at consumer-friendly terminals such as Singapore's Changi International Airport.[16]

In the final analysis, *Hubbing for Tourists* is a conceptual model of how aviation infrastructure, hub-heavy hometown airlines, and the closely related hotel

---

15 See Max Hirsh, *Airport Urbanism: Infrastructure and Mobility in Asia* (Minneapolis: University of Minnesota Press, 2016).
16 See John Bowen, *Low-Cost Carriers in Emerging Countries* (Amsterdam: Elsevier, 2019); Sven Gross and Michael Lück, eds, *The Low Cost Carrier Worldwide* (Abingdon Oxon, UK, 2013); Shelley Vishwajeet, The *IndiGo Story: Inside the Upstart that Redefined Indian Aviation* (Rupa Publications India Pvt. Ltd 2018).

sector, have contributed to tourism development throughout the Indo-Pacific Basin. It is meant as more of a conceptual argument about the structure and nature of Indo-Pacific commercial aviation rather than a chronological or comprehensive history of any of the individual sectors explored here (i.e., the city-states involved, respective airlines, or hotel properties or agglomerations).

Chapter 1
Before Dubai: The Shah, the Frank Lloyd Wright Foundation and Persian Gulf, 1967–1969

[T]his development could become one of the real wonders of the modern world and would create international respect and good will of the citizens of all countries for the manifest accomplishments of . . . Mohammed Reza Shah.
– Nezam Amery, Kamooneh-Khosrovi Architects (1968)[1]

We believe Dubai is unique. I like to think Dubai has forged its own path to prosperity using a home-grown formula . . . We have neither mimicked the paths that our energy-blessed neighbors have taken, nor have we copied the controlled economies of some other Arab states . . . We have created our own model.
– Sheikh Mohammed Bin Rashid Al Maktoum (2011)[2]

For many Iranians, Dubai's emergence as a global metropolis is imagined to have resulted, more specifically, from the displacement of Iranian modernity.
– Beyzad Sarmadi (2013)[3]

In the late 1960s,when Dubai was but a gritty port town and Jumeirah Beach had no (residential) Palm, the Frank Lloyd Wright Foundation dispatched retired U.S. Air Force officer David Fischer to the fringes of the Persian Gulf "to examine personally the Islands of Minoo (Salboukh) and Kish, and areas contiguous to both islands."[4] Fischer left Phoenix, Arizona, on September 24, 1967, traveling to New York and then to London by plane, before disembarking in Teheran. There, Fischer met with Iranian architect Nezam Amery and then spent two weeks consulting "government agencies, the oil consortium, NIOC, American Embassy economic sections, the American consulate, AID, United Nations, and private individuals" on the high-stakes mission entrusted to him.[5]

On October 12, Fischer caught a flight for the Persian Gulf delta town of Abadan. On the way, he stopped over in Isfahan to admire the Safavid's imperial maj-

---

1 Taliesin Associated Architects (cited hereafter as TAA), *Minoo and Kish Islands: Potentially Important Resorts for Modern Iran, Stage I,* (Scottsdale, AZ: The Frank Lloyd Wright Foundation, November 30, 1967), 28.

2 Pranay Gupte, *Dubai: The Making of a Megalopolis* (New York: Viking, 2011).11.

3 Bahzad Sarmadi, '"This Place Should Have Been Iran": Iranian Imaginings in/of Dubai', May 20, 2013, *Ajam Media Collective,* https://ajammc.com/2013/05/20/this-place-that-should-have-been-iran-iranian-imaginings-inof-dubai/, accessed August 28, 2018.

4 TAA, *Minoo and Kish Islands, Stage 1,* 6.

5 Ibid., 7.

https://doi.org/10.1515/9783111326641-002

esty, laid out in sumptuous gardens and spacious squares fronting a lavishly tiled mosque caparisoned with a *Timurid* dome. From there he traveled to Minoo Island—little more at the time than scattered fields of truck crops and livestock tended by farmers in the shadow of petroleum refineries. The Gulf itself left much to be desired—especially in view of his mission. "It is said that sharks infest these waters, to the extent that it is dangerous to swim," he candidly noted.[6] Fischer scouted out a local hotel, a supermarket, nightclubs, and "a floor show of variety with an international flavor, [including] Iranian and Arabian dancing, Swiss roller-skating, Japanese dancing, French dancing [troupes] and others."[7]

On October 14, Fischer flew south to Bandar Abbas, a city targeted by His Majesty, the Shah Mohammed Reza Pahlavi, for dramatic expansion. "There are new schools, a new stadium, a brand new theatre, newly asphalted roads," not to mention bustling bazaars featuring "products from all parts of the world."[8] Fischer also learned that an extended runway at the airport was in the works. The following day, he sailed to the nearby island of Kish, a desolate outpost inhabited by seven hundred people, and noteworthy only for pearl diving and a magnificently ruined seven-hundred-year-old palace. In surveying the island, he found the water supply and the soil suitable for expanded agriculture and development. He stayed over with the Hasheh village chief, borrowing his Range Rover to inspect the island before returning to the United States to report his findings to architect William Wesley Peters.[9]

What was the nature of Fischer's mission? Why such intense interest by an architectural firm in the hydrology and geology, not to mention the sociology, of some of the most inhospitable islands on the earth?

Fischer's nearly month long-journey preceded one of the most imaginative transcultural experiments in twentieth-century tourism planning. As part of a more ambitious undertaking by the Shah to situate Iran at the crossroads of global cultural and geopolitical prominence, the Iranian government commissioned the Frank Lloyd Wright Foundation—in cooperation with Iranian architect Nezam Amery—to develop a tourists' paradise on the Persian Gulf islands of Minoo and Kish just as Iranian economic prosperity reached its peak. As part of the Shah's initiative to modernize Iran's economy, also known as "the White Revolution," tourism played a dual role as visible proof of the nation's cosmopolitan aspirations.[10]

---

6 Ibid.
7 Ibid., 9.
8 Ibid., 11.
9 Ibid., 12–13.
10 Mohammad Reza Pahlavi, *The White Revolution*, second edition, (Teheran: Imperial Pahlavi Library, 1967). For an incisive analysis of the White Revolution, see James A. Bill, *The Eagle and*

Modern tourism development throughout the Middle East was not the exclusive domain of the Shah. Indeed, as Annabel Wharton has argued in her seminal study, *Building the Cold War: Hilton International Hotels and Modern Architecture*, U.S.-based entrepreneurs, backed by corporate power and the financial wherewithal of international aid organizations, planted modern hotels accented with local motifs throughout the region in cities including Istanbul, Cairo, Tel Aviv, and Teheran. Jet travel called for expanded accommodations. In Wharton's account, focused specifically on U.S.-based hotels chains in a global Cold War context, such oases of leisure were veritable "little Americas" transplanted into traditional, though progressive societies, anxious for the trappings of modernity.[11] Similarly, James E. Potter has argued in *A Room with a World View: 50 Years of InterContinental Hotels and its People, 1946–1959*, that Pan American Airways got into the act as well, contracting with local businessmen in places such as Beirut, to peddle American consumer culture and economic influence through leisure-oriented hotel projects.[12]

What these accounts leave out is the initiative taken by Middle Eastern governments, entrepreneurs, and architects to mold such initiatives to the needs of a *regional* clientele, including emerging Arab businessmen and their families, flush with petrodollars and eager to enjoy the pleasures of Western-style vacations in a context familiar to their preferences and aspirations. Indeed, this helps to explain, in the case of Wharton's study, why "local owners [of Hilton hotels in the Middle East] insisted on Modernity, sometimes in opposition to Hilton executives and designers, who desired more vernacular forms."[13] As the architects of the

---

the Lion: The Tragedy of American-Iranian Relations* (New Haven, CT: Yale University Press, 1988), 148.

11 Annabel Wharton, *Building the Cold War: Hilton International Hotels and Modern Architecture* (Chicago: University of Chicago Press, 2001). Wharton writes, "Like the great hotels of the colonial era in the Middle East, the postcolonial Hiltons embodied the cultural values of home veiled with references to the local. Interior design bore the burden of representing the host site . . . In Cairo and Istanbul ancient artifacts and indigenous crafts were deployed to provide Hilton interiors with their regional essence." (4) More recently, Begüm Adalet has examined leisure spaces as "built laboratories where [experts] could scale down problems of geopolitics and development to a manageable size and where they could test and cultivate modern subjectivities." See Begüm Adalet, *Hotels and Highways: The Construction of Modernization Theory in Cold War Turkey* (Palo Alto: Stanford University Press, 2018), 4.

12 James E. Potter, *A Room with a World View: 50 Years of InterContinental Hotels and its People, 1946–1996* (London: Weidenfeld & Nicolson, 1996), 8–12. Though an internally sponsored publication, Potter's book gives a fine, even-handed assessment of InterContinental's growth around the world.

13 Wharton, 6–7. The Istanbul Hilton, for example, lent its modernistic façade to Sedad Eldem, a Turkish architect who worked for Owings, Skidmore, and Merrill. See Welles Hangen, "The Tou-

study at hand elucidate, hypermodern architecture was an expression of power in the Middle East to a degree unrecognised by Western architects who hoped to straddle the visual semiotics of modernity, which from a "Western perspective" placed *equal* emphasis on the new and the traditional.

Furthermore, Middle Eastern and Persian Gulf investors and architects had taken it upon themselves by the 1950s to chart their own visions of tourism development, even in advance of the arrival of U.S. hotel chains. In Lebanon, for example, a new airport built outside of Beirut, quickly established itself as an entrepot between East and West in the early 1950s, drawing tourists to its own environs with the beauty of Byblos, nearby Damascus, and outdoor activities as diverse as skiing.[14] By the early 1960s, Beirut supplanted Cairo as the "must see" destination in the Middle East, with its pulsating nightclubs, easy visa requirements, and European cachet. Furthermore, regional demand for modern enclave developments propelled investors and developers farther afield, where the *wunderkind* prince Aga Khan IV underwrote the sprawling thirty-five-mile Smeralda Coast tourism enclave on the sparkling turquoise fringes of Sardinia—part of a multi-billion-dollar economic development project spearheaded by the Italian government.[15]

In the context of Iranian tourism development, the Taliesin Architects Association's vision for coastal Iran was only one part of broader proposals fired by the Shah's imagination. For his critics, these outsized leisure projects provided more revolutionary fodder for disgruntled Iranians. Under the Shah's direction, comprehensive plans to overhaul cultural development of historical *carvansarei*, the National Museum, and even the majesty of Isfahan—enveloped at the time in the

---

rists' Turkey and a New Hotel," *The New York Times*, March 13, 1955, XX29, *ProQuest Newspaper Database*.

14 Olivia Snaije explores the hubbing phenomenon in Lebanon in a recent article entitled, "My Airport: Beirut International Airport; Nostalgia and a Dream of Unity," *Popula*, https://popula.com/2018/12/17/my-airport-beirut-international-airport/, accessed January 14, 2023.

15 For an early assessment of Sardinia's post-World War II development needs, see Nick Mikos, "Development Works Drawing Sardinia Out of Isolation," *New York Times*, January 21, 1966, 78, *ProQuest Newspaper Database*. Mikos notes that the Aga Khan IV planned to develop "35 to 40 medium-size hotels, several hundred apartments and private villas on or near the beach . . . in the next 15 years at a cost of $240 million." Thomas W. Ennis followed up on Mikos's reporting a year later with his article, "Aga Khan's Brain Child in Mediterranean Is Growing Up," *New York Times*, July 16, 1967, 214, *ProQuest Newspaper Database*. Ennis estimated that the project would encompass some thirty-five miles of beachfront property with a development tab in excess of $300,000,000. Most importantly, the Costa Smeralda project remains today a viable enclave development.

exhaust of cars which choked the city's once serene streets— culminated in a pastiche of the old and new bedecked primarily for international visitors.[16]

While the planning documents for Kish and Minoo describe an ambitious vision for tourism, the geopolitical significance of the project also supports the argument that Persian Gulf states positioned themselves in the global imaginary as partners rather than pawns in an emerging neoliberal order. As Arang Keshavarzian has observed,

> in contrast to conventional accounts of globalisation that rest on ideas of modernisation, Westernisation, and liberalism, global processes need to be seen as multiple in origins. Rather than begin with the idea of [such spaces] as a Western-inspired form resulting from the application of the neoliberal programmatic, we need to understand them as indigenous responses [combining] local and global imperatives.[17]

Most striking is the similarity between what the architects elaborated and what a nearby emirate—Dubai—has become in the years since the plan's inception. While there is no direct connection between the Wright Foundation's designs for Minoo and Kish Islands and what has materialised in Dubai (and nearby Persian Gulf citystates including Bahrain, Doha, and Abu Dhabi), the similarities speak to common aspirations at what was once the very crossroads of human civilisation.[18]

This chapter examines the transcultural vision of Amery and Peters in conceptualising touristscapes for Minoo and Kish Islands. It emphasizes the hybrid approach of the architects in appealing at one and the same time to Arab and Western tourists, the staggering environmental costs of such an undertaking, and the role that contemporary tourism in Beirut played in shaping their vision for a new center of tourism closer to the Arabian Peninsula. While the chapter makes no attempt to link plans for the Iranian periphery to Dubai (one of the seven emi-

---

16 In 1965, the Shah personally received UNESCO officials invited to Teheran to assess the country's cultural patrimony in efforts to promote tourism there. See UNESCO, *Iran: Preservation et Mise en Valeur du Patrimoine Monumental de L'Iran en Liaison avec le Developpement du Tourisme dans ce Pays.* Paris: UNESCO, June 1967. Nearly a decade later the Shah also oversaw a bold initiative to redesign downtown Teheran, a plan known as the *Shahestan Pahlavi.* See Deyan Sudjic, *The Edifice Complex: How the Rich and Powerful Shape the World* (New York: The Penguin Press, 2005), 177–184.

17 Arang Keshavarzian, "Geopolitics and the Genealogy of Free Trade Zones in the Persian Gulf," *Geopolitics*, 15 (2010), 266.

18 For specific studies of architectural trends in Dubai and the Gulf, see Heinfried Tacke, Sabina Marreiros, ed, and Suzanne Kirkbright, trans., *Dubai: Architecture & Design* (Cologne, Germany: Daab, 2006); Archis, AMO, and Moutamarat, present an early overview of architectural and design achievements in *Al Manakh* (Amsterdam: Archis, 2007); Todd Reisz edited a follow-up collection to the first volume in *Al Manakh: Continued* (Amsterdam: Archis, 2010); More recently, see Yassar Elsheshtawy's *Dubai: Behind an Urban Spectacle* (London: Routledge, 2013).

rates which comprise the United Arab Emirates—known prior to its independence as the Trucial States), it does suggest that similar models—in their appeal to multiple audiences, as well as their position at the crossroads of continental convergence between Africa, Asia, and the Middle East—, capitalised on then circulating ideas of a long-term strategic vision to reorient the economic and cultural landscapes of the twenty-first century Persian Gulf.

The chapter concludes with a brief introduction to the disruptive model Dubai posed in aviation and tourism circles well before the incorporation of Emirates Airlines in 1985. Based on documents from the British National Archives, it suggests that plans for Dubai's airport and the actions of state leadership anticipated an "open skies" model at the expense of commercial aviation cooperation with the rest of the Emirates. By the early 1970s, British diplomats reckoned with the fact that Dubai might not only serve as a strategic aviation entrepot, but also could conceivably launch a long-haul airline servicing a promising tourist destination.

***

In the late 1960s, Iran boasted a robust economy and its leaders aspired to greater global prominence beyond the Persian Gulf. Recent initiatives encompassed in the Shah's 1963 White Revolution, including land reform, liberalisation of women's rights, and administrative modernisation, positioned the country to make the jump from an emerging, largely agrarian economy to a developed nation. Iran industrialised at a rapid pace and urban growth mirrored this structural shift. Outside of Teheran, no region grew faster than Khuzestan, the province where Abadan (and adjacent Minoo Island) was located. Population there doubled largely because of migration from other parts of the country. Iran's major port, Khorramshahr, anchored regional trade there, and oil refineries employed upwards of 45,000 Iranians and foreigners. Tourism development and the designation of Khorramshahr as a free port represented the next steps in regional development and diversification.

Under the direction of the Ministry of the Interior, William Wesley Peters and Nezam Amery set to work designing a comprehensive plan for tourism development on Kish and Minoo Islands.[19] The pair charged the Iranian government thirty thousand dollars for a two-stage report, the first of which included a conceptual vision of the tourist attractions and infrastructure on the two islands. The second stage filled two dossiers with a meticulous assessment of natural conditions on the

---

19 Of the relationship between the two architects and the government of Iran, Amery wrote, "Wes Peters was a great friend of mine and we had already done the Palace of Princess Shams together so I introduced Taliesin to participate in this project." (e-mail to the author, December 18, 2007).

islands, the resources necessary to bring the project to completion, and comparative views of competing projects in the Middle East and around the world. When the Iranian government scaled back the timeline for the project in 1968, the architects prepared a modified plan to match a longer period of development.

Neither Frank Lloyd Wright, nor Peters, his heir apparent and son-in-law, was a stranger to the hospitality industry. Wright himself designed six hotels during his life. The most famous of these, the Imperial Hotel in Tokyo, dazzled with a cross-cultural design that boldly celebrated the vertical tendencies of American architecture as well as the more discreet and subtle touches of Japanese inns.[20] Peters not only completed the complex engineering designs that made many of his father-in-law's most famous projects a reality, including the home at Fallingwater and the Guggenheim Museum, but himself designed the Snowflake Motel, a sprawling complex of "six V-shaped units surrounded by a driveway and connected by a geodesic dome" in southern Michigan.[21]

Several years after completion of the Snowflake Motel, the Shah of Iran's sister, Shams Pahlavi, commission Peters to build an opulent palace near Teheran. Peters apprenticed Amery to assist with that project, which was completed in 1966—just a year before plans for Minoo and Kish were conceived. In the early 1990s, Peters reflected on the design process for the futuristic home, which boasted 50,000 square feet and "culminated in the Princess's bedroom with a spiral ziggurat." In an interview with Ken Burns for his documentary on Frank Lloyd Wright, Peters remembered, "Before [the princess] told us about what she wanted she sent me to see some of the historic spots of Iran . . . Nezam Amery and I went down to see Persepolis and Isfahan and a number of other beautiful sites and cities in Iran."[22] The domes and ziggurat of the Pearl Palace, as well as the imaginative layout of the Snowflake Motel, all contributed to the conceptual design for the cityscapes on Minoo and Kish Islands.

Spatially, the architects considered the relative absence of development on the islands advantageous. The state could assert its absolute fiat over these sparsely populated areas in one of two ways: either by exercising imminent domain or endowing existing landowners with a stake in the project's potential profits. Political scientist Waleed Hazbun refers to such a process as the 'territorialization' or repur-

---

20 See discussion of the hotel in Berry Bergdoll and Jennifer Gray, eds., *Frank Lloyd Wright: Unpacking the Archive* (New York: Museum of Modern Art, 2017).

21 *South Bend Tribune*, "Snowflake Melts into History," January 26, 2006, http://articles.south bendtribune.com/2006-01-26/news/26921686_1_snowflake-architectural-community-architects-split, accessed August 28, 2018.

22 Nima Kasraie, "Spiraling into Oblivion," *The Iranian*, June 4, 2004, https://iranian.com/2004/06/04/spiraling-into-oblivion/, accessed August 27, 2018.

posing of a space intended for tourism development.[23] This trend gained momentum in the 1960s for a variety of reasons, both economic and cultural. Economically, total state control over spaces like Minoo in Iran or Cancun Island in Mexico enabled the state to manage all levels of development (as well as reap all potential rewards from real estate sales) in a designated area. Culturally, the creation of satellite cities in tourist zones separated hotel and leisure enclaves from adjacent cultural attractions (such as Persepolis and Shiraz, for example) much in the same way that the creation of La Défense in Paris was intended to prohibit skyscrapers from sitting cheek-by-jowl with historical structures near the center of the French capital.[24]

Development on the periphery also targeted an untapped market in the region where "with the exception of major and/or capital cities, the source countries offer their own citizens little in the way of recreation or resort areas."[25] As the eventual reluctance of the Iranian government to fully fund the concept proved, peripheral tourism enclaves required significant investment and posed a substantial financial risk. In fact, it was as likely that tourism mega-projects devised by foreign aid organizations, emerging governments, and international consultants, between 1960 and 1985, failed than succeeded.[26]

Be that as it may, the two islands occupied a prime location ripe for development. Such had been the case for millennia. As Keshavarzian noted, "In earlier centuries commerce and distribution of credit was embedded in a rich social mi-

---

23 Hazbun, *Beaches, Ruins, Resorts*, 216–220.

24 The idea of creating a satellite community was very much in the forefront of Amery and Peters' design. Amery observed, "Initially it was thought it would be a good idea to build a centre from which the tourists could visit the archeological sites of the area but gradually our thoughts developed into the idea of making the site itself a tourists' attraction as well as a place to stay and visit the surrounding sites. Naturally being Persian and brought up on the Magic, Myths, and Literature of Persia, it was immediately obvious to me in the direction we would go, i.e. a magical mythological wonderland." (e-mail to the author, December 18, 2007).

25 TAA, *The Islands of Minoo and Kish: Tourist Feasibility Analysis and Report for the Ministry of the Interior, Stage 2*, Volume 2 (Scottsdale, AZ: The Frank Lloyd Wright Foundation, August 1968), I-1.

26 Developments in the Caribbean provide several proofs of the hit-and-miss realities of large-scale enclave tourism development. For example, even though FONATUR (the Mexican government's beachside tourism development arm) showcases Cancun and Los Cabos as economic success stories, the Mexican government suffered setbacks at Ixtapa, Bahias de Huatulco, and Loreto – all enclave tourism developments which failed to attract significant tourist interest. The Ixtapa project had been underwritten by loans from the World Bank. It might also be argued that a similar development at Puerto Plata, Dominican Republic (also underwritten by the World Bank), never lived up to expectations, while private projects on the east coast of Hispañiola, at today's Punta Cana and Casa de Campo (near Santo Domingo), had more enduring economic success.

lieu of kinship networks, religious lineages and imperial processes that made the waterway and littoral lands a bridge between Asia, Africa, and Europe."[27]

The island of Minoo sat adjacent one of the world's most productive oil producing areas. As such, it was a magnet for Arab businessmen and tourists, as well as the occasional European or North American. In contrast, Kish Island presented a topographical *tabula rasa* upon which innovative transportation strategies could be engineered. From the outset, Amery and Peters envisioned an international airline hub and transit zone there. "With the development of landing facilities for SST/Jumbojet aircraft, [Kish] possesses the geographical location for a major stopover point for air travel," their report noted.[28] As such, "Western Europeans may well show a preference for Kish . . . since its location is useful to continuing journeys to Japan and Australia, Australasia and the Western Hemisphere."[29] In this aspect, Kish followed Dubai's lead, which opened an international airport in the late 1950s under an "open skies" policy.[30]

Whether the Shah planted this idea with the pair of architects, or they influenced his thinking, is hard to tell. Yet years later, British bureaucrat Michael Heseltine reported the gist of a conversation with the Shah in 1972 respective of the nation's place in the world. "It was understood that the Shah of Iran had a vision of Teheran as a staging post between West and East," Heseltine remembered. "He saw [the] Concorde as an important part of the process, if [Teheran] was seen as a major stopover on its journey both ways."[31]

Nearly sixty years on, such thinking for Kish was not far from the long-term vision for Dubai. It would take some time—formally marked, perhaps, by the incorporation of Emirates Airlines in 1985—before Dubai attained the status of both a strategic aviation and tourism hub. Yet, as both Keshavarzian and Hazbun have argued, this strategy boldly challenged entrenched power players in region, be

---

27 Keshavarzian, 270.

28 TAA, *The Islands of Minoo and Kish: Tourist Feasibility Analysis and Report for the Ministry of the Interior, Stage 2*, Volume 1 (Scottsdale, AZ: The Frank Lloyd Wright Foundation, August 1968), 95.

29 TAA, *The Islands of Minoo and Kish, Stage 2*, Volume 2: I-1.

30 Hazbun, 206.

31 Michael Heseltine, *Life in the Jungle: My Autobiography* (Hodder and Stoughton Ltd, 2000) as cited in "The Late Shah of Iran's Vision," *Imperial Iran*, August 19, 2004, http://imperialiran.blog spot.com/2004/08/late-shah-of-irans-vision.html, accessed August 27, 2018. Neither the Taliesin architects, nor the Shah, were alone in their awareness of the strategic importance of the Persian Gulf corridor as an aviation crossroads. In terms of Gulf competition for preeminence as a stopover point, journalist Marvin Howe wrote, "Because of its central position on the Persian Gulf, Bahrain is one of the principle stopovers on air routes between Europe and the Far East, and picked up a good deal of Beirut's traffic during the Lebanese civil war. See Howe, "Bahrain Concerned Over Mainland Link," *The New York Times*, July 24, 1977, 9 *ProQuest Newspaper Database*.

they more recent Cold War overtures from the United States or the Soviet Union, or enduring post-colonial ties (in the case of European nations, namely Great Britain).[32]

Peters and Amery conceived of such a function for Kish in geographic terms familiar to them at the time. "It seems likely, therefore, that a pattern of tourist facility [on Kish] not unlike that of Beirut or Hong Kong would best fit the projections for the Island," they noted, "that is, a transient kind of trade, [with airline passengers] visiting from 2 to 5 nights with heavy emphasis on stop over and short vacation travel."[33] Tourist friendly transit regulations allowed passengers up to a seventy-two hour stay in Iran without the need for a visa. The national airline, Iran Air, whose footprint in the Middle East and Europe had rapidly grown in the 1960s with the Shah's encouragement, offered liberal discounts (between 25 and 40 percent) to passengers visiting the country.

Aviation consultant and scholar, John Kasarda, has written about the emergence of cities anchored around airports in the late twentieth century. He calls such a city an "aerotropolis," citing Dubai as a prime example. "Dubai discovered after 9/11 that its greatest asset wasn't oil but *geography*," he and Greg Lindsey wrote, "defined not by the contours of any map but by the flying times of modern airlines. Living within four thousand miles of the emirate, less than an eight-hour flight away, are 3.5 billion people, more than half the world's population."[34] Kasarda and Lindsey also highlighted a social reality that Peters and Amery figured into their model of Kish and Minoo: that it would offer amenities not only for the wealthy, but also the emerging classes both in the Middle East, Africa, and Asia. Through the instrumentality of Emirates Airlines, Kasarda and Lindsay have written, Dubai achieved what Peters and Amery could only theorise. Speaking of the billions living within an eight-hour reach of Dubai, Kasarda and Lindsey observed, "they're the ones learning the world is flat – billions of the most deprived, the most striving, and the most nouveau of the world's riche, all dressed up with

---

32 In his article Keshavarzian notes, "This analysis illustrates that the Middle East is neither absent from the process of, nor does it simply respond passively and reactively to this complex process. Free trade zones are an example of local strategies working in consort with international processes to fashion new forms of economic and political interconnectedness" (263). Similarly, Hazbun argues that more recently, "What is different . . . is that because of expanded regional flows of tourists and capital, new patterns of consumption and entrepreneurial strategies, and better marketing, many Arab states and firms have increased their ability to influence the size and pattern of these transnational flows. As a result, they can now promote economic reterritorialization not only by controlling local tourism spaces but also by influencing networks at the global, but more so the regional, level" (196).

33 TAA, *The Islands of Minoo and Kish, Stage 2*, Volume 2: G-8.

34 Kasarda and Lindsey, *Aerotropolis*, 295.

no place to go . . . With no hinterland of its own, this desert enclave scrambled to offer the only hub they'll ever need – their London, Miami, Las Vegas, and Singapore all wrapped up in one – the navel of the middle of the world."[35]

State planners believed that Iran Air's expanding network would attract a larger pool of tourists during the 1970s and 1980s. Iran Air boasted a fledgling fleet of 15 planes, including one Boeing 727. Domestic routes crisscrossed the country from east to west. International connections included destinations from Asia to Western Europe, including Kabul, Karachi, Kuwait, Istanbul, Rome, Paris, London, and Frankfurt. By the late 1970s the network included service to New York twice a week. Iran Air carried a mere 11,000 travelers in 1953. By 1965 that number surged to 310,000 passengers.[36] With an estimated one million tourists targeted to visit both Kish and Minoo by 1985, Iran Air, as well as international carriers, played an essential role in the global reach of Iran's leisure industry. As one of its 1969 ads featuring the word "Tehran" spinning towards the reader (in the motion of a hub) boasted, "When flying from Europe to the Near and Middle East, go via Tehran. From this town of nearly 3 million inhabitants IRAN AIR can offer comprehensive regional to all the main economic and cultural centres."[37]

The development plan for Kish and Minoo targeted a sophisticated mix of regional (Arab and Islamic visitors) and international tourists (Europeans and North Americans), whose joint presence would establish a rich mosaic of affluent visitors. Arab tourists constituted the largest and most lucrative market for development on the Iranian islands. Market research revealed that this segment, which generally traveled as families or extended kin networks, also spent more money per person and stayed for longer periods of time (i.e., 10–12 days) than their Western counterparts. Given the considerable flows of businessmen and their families to the Persian Gulf delta in Abadan, adjacent Minoo Island would cater to tourists of this ilk. Peters and Amery outlined plans for a larger number of condominiums and apartments as part of the infrastructure than would have been the case if the enclaves appealed exclusively to Western tourists, who were inclined towards staying in hotels for shorter periods of time (i.e., 3–5 days).

Nevertheless, Iranian authorities had their work cut out for them. "All these [Arab tourists] seek that which is more luxurious than the conditions under which they themselves live, and will look for the unusual, the historic, the ancient and the modern, in the resort areas which they will patronize," the report observed. "To get these people to an area once is less difficult than to get them to

---

35 Ibid.

36 TAA, *Minoo and Kish Islands, Stage I*, 109–111.

37 The author located this ad on e-Bay during the final stages of editing and cites it here although the URL was not permanent.

repeat, say, once in every three or five years, and it demands a marketing program of continuing but varying attraction."[38] For those that have since been to Dubai, such demands parallel the constant pursuit of the largest and latest innovations. This is as much for regional appeal as it is intended for a Western audience.

Peters and Amery also made distinctions between Arab, North American, and European tourist preferences. American and Canadian tourists were likely to visit Minoo Island only while it was considered a "bucket-list" destination. Lastly, although European tourists did not demand the same caliber of amenities as their Arab and North American counterparts, they expected to be able to visit cultural or historical attractions near hotels and "satellite cities," much like those described above. For example, visitors to Kish Island might fly to Shiraz and take a tour of nearby Persepolis before returning to Kish for several days on the beach.[39]

Amery and Peters' ambitious project was inconceivable without several technological innovations, chief among them air conditioning, an ample water supply, and wind abatements. As with many large-scale tourism projects planned prior to the OPEC oil crunch, the architects were confident that human ingenuity would triumph and even transform the desert into a new paradise. "In the refreshing and tranquil environment of modern resorts carefully designed to respect the tradition and unique environment of their location," William Wesley Peters airily wrote, "the 20th Century visitor may steep himself in the flow of civilization from the Garden of Eden to the present while enjoying the comforts and luxury of modern Iran."[40] Thus, the project's design depended entirely on manipulation of the environment and the sustainability of scarce resources. Almost all activities, save horse racing, soccer, swimming, and water-skiing, would take place indoors with the benefit of massive cooling systems (sixty years later the Qataris would host the 2022 World Cup in air-conditioned *outdoor* stadiums). "A high temperature range throughout most of the year emphasizes the need that considerable thought be given to year-round temperature control and total air conditioning within all the proposed facilities," the report advised.[41] The hotels would be built around central courtyards that could be opened to cooler climes during winter months. Shopping centers, restaurants, and theaters would be situated close to hotels and connected by cooled passageways. Extravagant fountains would also aid in lowering temperatures as well as creating a Babylonian redux. Abundant water would:

---

38 TAA, *The Islands of Minoo and Kish, Stage 2*, Volume 2: I-1.

39 Ibid., volume 2: I-1–2.

40 TAA, *The Islands of Minoo and Kish*, Stage 1, Preface.

41 TAA, *The Islands of Minoo and Kish, Stage 2*, volume 2: I-28.

make possible extensive landscaping utilizing numerous water courses, fountains, pools, reflective basins, gardens, flowers, trees and foliage that would display in a modern form the ancient tradition of the Persian garden.

Receding terraces, overhanging balconies, covered promenades, terraced patios, could all be enriched by the colorful charm of growing vines, blossoming shrubs, and garden flowers.[42]

Wind abatement constituted a third obstacle. The three-thousand-seat amphitheater required "a screening device to protect the actors and audience from winds that have velocities of more than 10 miles per hour, which could disturb good syllabic articulation at the rear of the audience."[43] Even the schools, constructed for the anticipated fifteen thousand residents of the completed tourist bubble, would benefit from the massive expenditure on environmental modifications. As the report indicated, "the [schools'] layout and orientation should aim to secure protection from intense sun by the creation of shade areas, shutting out heat and dust-laden winds, generous provision of planting, water fountains, plunge pools and good natural and artificial ventilation."[44]

The architects compared the potential needs for water and power on Kish and Minoo with benchmarks for power and water usage in Miami, Beirut, and Houston's Astrodome. They believed that existing systems on Minoo and new infrastructure at Kish could provide the necessary power and 140 to 160 gallons of water per day for each tourist. Surprisingly, no mention was made of desalinisation as an alternative to draining limited groundwater reserves on Kish Island.

Confident in their ability to surmount the obstacles posed by nature, Amery and Peters envisioned the two islands as peerless tourist destinations. Amery surely did not exaggerate when he wrote to Iran's Minister of the Interior: "The consultants believe that the potential accomplishment and gain . . . by way of the simultaneous . . . development of the islands of Minoo, Iran and Kish is enormous, and . . . offers an opportunity not enjoyed by any other country or area with which the Consultants have had experience."[45]

Minoo Island would be the more extensively developed of the two islands. This was due in part to the ready availability of electricity, produced by a dam recently built upriver, as well as water from Abadan's existing reserves. Building a "beachside" resort required illusions as attractive as desert mirages. As David Fischer had learned on his visit to Iran, both Minoo and Kish Island were home to

---

42 Ibid.

43 TAA, *The Islands of Minoo and Kish, Stage 2*, volume 1: 74.

44 Ibid., volume 1: 71.

45 Ibid., volume 1: 9.

shark populations that made swimming precarious at best. The passage of large ships also threatened to inundate the islands. To prevent either eventuality, the architects proposed that a promenade, built above a barrier six-feet high and twenty-four feet wide, encircle the island. Recessed swimming areas protected swimmers from dangers lurking in the water.[46]

A luxury hotel, apartment buildings, and condominiums with an adjacent motel for short-stay visitors would serve as the centerpiece for island activities. As noted above, the rooms would surround an inner court, lobby, and "grand hall," protected from "sun, rain, winds and dust storms."[47] Large glass windows would afford expansive views of the Gulf. Shopping centers, a three thousand-seat amphitheater, a concert hall (from eight hundred to three thousand seats), and movie theaters would project "a glittering skyline of spectacular interest to passing ships and air traffic for miles around."[48] The bazaars would cater to travelers on a new aerial Silk Road, "completely air conditioned, ventilated, and lit, but arranged so that the charm and intimate quality of the old bazaar is retained."[49] What would not be lost, the designers hoped, was the mystique of ancient marketplaces, as well as clues of Iran's modernising tendencies. "Visiting tourists," the report noted,

> would experience, in a modern surrounding, the same intriguing lure that has attracted them to the bazaar of old. They would proceed from place to place to find small shops and merchants dealing in jewelry, hand crafted products [from all regions of the country], art works and fabrics, as well as to larger rooms and display areas showing the products of modern Iranian industry and science.[50]

The illusion of abundant water would be everywhere. "An arrangement of gardens with fountain jets in reflective basins would grace the interior space" of the multi-purpose complex.[51] If business became too brisk, however, waterfalls and gardens could be "removed to make way for other displays if desired."[52] Illusion rivaled water and glass as the chief vehicle for appeal. *Desert baroque* might have been the most apt description of the movie theatre, appointed with "a festive atmosphere within the precincts of the building like that of a fairyland come alive."[53] Passage to the viewing areas themselves would be an odyssey, with

---

46 Ibid., volume 1: 47.
47 Ibid., volume 1: 51.
48 Ibid., volume 1: 52.
49 Ibid., volume 1: 55.
50 Ibid.
51 Ibid., volume 1: 52.
52 Ibid.
53 Ibid., volume 1: 56.

movie-goers transiting "spacious lobbies gaily illuminated with discreetly located colored lighting; suitably bedecked with planting, flowers, and fountains, subtle placement of mirrors to reflect the to and fro movement of patrons through the spaces, plush carpets, abstract chandeliers and illuminated ceilings."[54]

Contrasts were an important part of the modern, global imaginary, not solely in temporal terms (providing for the old and the new), but also in the realm of attractions. The designers spent extensive time exploring the requirements for a three-thousand-seat auditorium that could be converted to different configurations for different types of events, including those of a conscientious citizenry. In addition to conventions, conferences, and grand balls, symposia might elevate the tenor of discussion amongst visitors to the island. "Speakers could . . . be selected who will most contribute to the better understanding of man, geography, space, science, art and national customs of the world," Peters and Amery recommended.[55]

On the other hand, the *piece de resistance* was a Persian-themed amusement park, whose description bears citing in whole. "With attention to the great success of Disneyland in the United States," they began:

> and going far beyond it in poetry, glamour, and imagination, this feature could form a year-round demonstration to Iran's citizens and visitors of the glory of the country's past and the greatness of its future.

> In the hands of an understanding Architect-Designer, this garden park could become in itself a work of art of such significance and quality as to attract interest and attention throughout the world and reflect understanding and appreciation of the signal achievements of Persia in the past and present.

> Most importantly the park should be essentially a place where the visitor is regaled with enjoyment; where the interplay of color, sculptural forms, rich materials, gardens, fountains, and flowers produce an overwhelming feeling of joy and beauty, a new Garden of Eden near the fabled location of the ancient one.[56]

---

54 Ibid.

55 Ibid., volume 1: 52.

56 Ibid., volume 1: 73. In a 2007 e-mail exchange with the author, Nezam Amery noted: "I had envisioned a Persian wonderland with mythological characters from the history of Persia for example, Rostam and Zohrab and other great heroes of the past. Of course there would have been hotels but everything would have had the character of Persia. At the time we were considering Kish and Minoo there was a renaissance in Iran and appreciation of everything Persian. There was a great feeling of national pride unlike ten years earlier when only the West was appreciated in Iran. The country had moved on and the middle classes wanted Persian artifacts and indeed Persian designs had thrived. We were creating the mythological history of Persia in a totally imaginary and creative status not a copy of anything which exists." (e-mail to the author, December 18, 2007).

Even more astonishing, the designers believed that such a complex could be built on *both* islands! Symbolically, the Persian park epitomised the transcultural nature of the tourism bubble itself. Its local motifs played to both a regional and international audience; on the one hand for its familiarity embodied in a venue suitable for family audiences, and on the other a "must-see" destination for an international jet-set looking for the latest exotic attraction. In this respect, what Waleed Hazbun has written about Dubai might also have been said about the design for the Iranian islands; to wit, "rather than viewing [Minoo and Kish's] cosmopolitan culture as a product of placelessness and cultural deterritorialization, unmoored from a territorial source of identity, it is useful to consider how the architecture of its urban form sustains a mosaic of connected cultures attached to multiple locations."[57]

Although the tourism infrastructure designated for Kish and Minoo varied mainly in scale and cost ($88 million dollars vs. $256 million, respectively), the climate, tourist-mix and the length of their expected stay differentiated the design for the former island. Initially, start-up costs were proportionally higher for Kish given the lack of sustainable resources. Though most visitors would hail from adjacent Persian Gulf states and the Arabian Peninsula, transit stays of between three and five days mitigated the need for attractions as extensive as those envisioned for Minoo. With its higher temperatures, moreover, vacationing guests would likely opt for winter visits rather than during the brutal summers when temperatures averaged 120-degrees Fahrenheit.

Finally, if the Iranian government decided to build a major airport hub on this converging nexus of Asia, Europe, and Africa, many guests would do nothing more than shop duty free prior to departing Kish Island. Tax breaks for airlines and retailers would swell the number of flights and the girth of visitors' luggage respectively. The architects cited Puerto Rico as an analogue for shopping tourism. Transiting tourists spent approximately ten percent of their total expenditures on merchandise while in the American commonwealth territory. "This figure can easily be a great deal larger for Minoo and Kish depending on the extensiveness and product range and variety of bazaars," Amery and Peters confidently boasted—as if describing the duty-free havens that would fill Dubai's airport less than two decades later. "The public knowledge of very large and fine bazaars would depend on advertising and tourist recommendations, which would greatly enhance their shopping potential."[58]

---

57 Hazbun, 223.
58 TAA, *Minoo and Kish Islands*, Stage 2, volume 2: J-2.

As with any professional proposal, Amery and Peters pointed to existing tourist developments to justify their fantastic vision. To be sure, there were no enclave developments indigenous to the region at the time (Costa Smeralda was located on the Italian coast of Sardinia) that paralleled their proposal, but they cited Beirut, the most popular Arab destination at the time, as a comparable analogue—or one that they aspired to supplant.[59] This point of comparison says something about the ambitious nature of plans for Kish and Minoo. This would not be simply another tourist destination, but *the* leisure site in the Persian Gulf and Arab tourist imaginary.

Simply put, during the 1950s and 1960s, Beirut was king of the Middle Eastern tourist world: not solely for its myriad attractions, but also as the first strategic, post-World War II stopover point between East and West – a position that Bahrain and Kish aspired to in the 1960s and 1970s, and Dubai would achieve in the 1990s (Istanbul, Abu Dhabi, and Doha have extended this pursuit into the twenty-first century). As Hal Lehrman noted in his *New York Times* article of April 8, 1956, "Beirut . . . owes much of its new importance to the expanding needs of global aviation. Its excellent position as a midpoint between West and East has attracted one international airline after another – among them Pan American, B.O.A.C., Air France, Swissair, and Air India."[60]

But air access alone did not attract tourists from East and West simply to fly elsewhere. The French cachet that gilt conversations in nightclubs and cafes had long attracted Europeans and Arabs to this modern oasis. The beach was another attraction. "Beirut's seacoast is fringed on the south by long beaches of golden-colored sand, and on the north by banana and orange groves," wrote journalist Ruth Warren.[61] The sons of sheikhs ducked in and out of Beirut's watering holes,

---

59 In addition to the Costa Smeralda project, Peters and Amery would likely have been aware of Europe's greatest project created on a blank geographic slate: Grande Motte – whose history is treated by Ellen Furlough and Rosemary Wakeman in "La Grande Motte: Regional Development, Tourism, and the State," in *Being Elsewhere: Tourism, Consumer Culture, and Identity in Modern Europe and North America* (Ann Arbor: University of Michigan Press, 2001), 348–372.

60 Hal Lehrman, "Boot Town of the Mid-East," *New York Times*, April 8, 1956, SM 110, *ProQuest Newspaper Database*. C.R. Squire made a similar point a year earlier in "Levantine Air Links: Visiting the Arab States is Easy and Cheap," *The New York Times*, May 15, 1955, X47, *ProQuest Newspaper Database*.

61 Ruth Warren, "Sight-Seeing In and Around Lebanon," *New York Times*, February 11, 1951, 107, *Proquest Newspaper Database*. The Lebanese government promoted tourism development as an extension of modernisation, especially through its National Council of Tourism. See Elie Adib Salem, *Modernization without Revolution: Lebanon's Experience* (Bloomington: Indiana University Press, 1973), 98–99.

where they "sip[ped] alcoholic drinks without Koranic scandal."[62] By the 1960s, it had supplanted Cairo as the go-to destination in the Middle East.[63] It's cosmopolitan vibe contrasted wildly with the more staid façade of Damascus.[64] Recurrent unrest, however, prevented Beirut from consolidating its hold on its regional preeminence as a tourist destination.[65]

According to the Lebanese tourist authorities whom Peters and Amery consulted, 1.6 million tourists and other travelers (which included travelers from nearby Syria and Jordan) visited Beirut in 1964. Among them were approximately 500,000 European tourists. Lebanese officials estimated that one million tourists alone would arrive in the capital by 1968; two million by 1975; and four million by 1980. They observed that Arab families headed to quiet mountain hotels or rented spacious apartments in the city for periods from ten to twelve days. "For Arab tourists," the Wright Foundation report noted, "the attractions [of Beirut] mentioned first were distractions [i.e. shopping] and amusements, modern Lebanon (its development, Beirut's expansion, its buildings, the impressive progress), and the mountains of Lebanon."[66] In contrast, nearly all Europeans and North Americans stayed in the city on average four days. In contrast to their Arab counterparts, these "westerners" preferred "archeological and historic sights, modern Lebanon, and it is agreeable to say, the Lebanese themselves, their hospitality, their kindness and courtesy."[67]

---

62 Lehrman.

63 See Thomas F. Brady, "Beirut Leads Area's Book: Lebanese Metropolis Supplants the City on the Nile as the Favored Tourist Capital of the Middle East," *The New York Times*, March 6, 1966, XX49, *ProQuest Newspaper Database*.

64 Author Félix Martí-Ibáñez wrote of his visit to Beirut in 1966. He reveled in the whiff of Paris he found in the café's and compared its racial diversity with that of its Near Eastern counterparts to the north and south. See Martí-Ibáñez, *Journey Around Myself: Impressions and tales of Travels Around the World: Japan, Hong Kong, Macao, Bangkok, Angkor, Lebanon* (New York: Clarkson N. Potter, Inc., 1966), 373–378.

65 For discussions of Beirut's resurgence as a tourist hub after conflict in the 1950s and 1960s, see Richard F. Hunt, "Lebanon Hopeful: She seeks Early Return of tourists After Year of Siege and Shooting," *The New York Times*, March 1, 1959, *ProQuest Newspaper Database*; and, Dana Adams Schmidt, "Back in Business: Lebanon Woos Tourists as Unrest Subsides," *The New York Times*, February 25, 1962, XX40, *ProQuest Newspaper Database*. More recently, devotees of Anthony Bourdain's various travel shows will recall his violence-marred visit to Beirut that ended in a military evacuation from the troubled city. See Kim Ghattas, "How Lebanon Transformed Anthony Bourdain," *The Atlantic*, June 9, 2018, https://www.theatlantic.com/international/archive/2018/06/how-lebanon-transformed-anthony-bourdain/562484/, accessed November 5, 2018.

66 TAA, *Minoo and Kish Islands*, Stage 1, 183.

67 Ibid., 184.

In retrospect, these were prescient insights. The Arab preference for hyper-modernity suggests that—in a different urban iteration—Dubai's outsize appearance has as much to do with attracting regional visitors as luring Westerners to the region.[68] The same might be said for other hybrid cityscapes like Shanghai, whose Pudong district sends subtly distinctive messages to different constituencies: to locals it affirms the resurgence of Chinese power vis-à-vis historical Western imperialism. At the same time, the soaring architecture stakes a claim to global prominence.[69]

Amery and Peters not only hoped that Iranians could learn from the Lebanese, but also profit from their success in a highly competitive market. "We are also studying how to increase contacts by tourists with the Lebanese." They identified a new type of traveler, the "vacationer clientele," or tourists interested primarily by "the sun and the sea."[70] Regional synergies would benefit both tourism poles, with largely uninhabited islands in Iran much more conducive to large-scale resort development than Beirut. Catering to the "vacationer clientele" became the great pursuit of tourism developers around the world – in places like Thailand, the Dominican Republic, Mexico, the Balearic Islands, and Turkey.

***

When David Fischer returned from his three-week mission to the Persian Gulf in 1967, his report anticipated many of the advantages that Dubai has exploited in its own development of tourism. In fact, the story of Dubai's rise to aviation prominence began prior to Fischer's junket to the Persian Gulf but only blossomed decades later. Beginning in 1957, Sheikh Rashed bin Saeed Al-Maktoum (1912–1990) repeatedly importuned the British government to build a commercial airfield at Dubai. The British reminded the Sheikh that such an establishment already existed in neighboring Sharjah, some twenty miles away. In 1959, Dubai raised the stakes, informing Her Majesty's Government that an Arab empresario had agreed

---

68 Sandy Isenstadt and Kishwar Rivzi, "Introduction: Modern Architecture and the Middle East: The Burden of Representation," in *Modernism and the Middle East: Architecture and Politics in the Twentieth-Century*, Sandy Isenstadt and Kishwar Rivzi, eds. (Seattle: University of Washington Press, 2008). The authors note that "Questions of architectural representation remained inescapable in the Middle East because the primary issue was to make modernity and independence manifest, to visibly demonstrate with material form claims of political parity with former colonial and hegemonic powers" (22).
69 See Kirsten Day, "The Shanghai Paradox," The Mediated City Conference, Woodbury University, April 1–3, 2014, http://architecturemps.com/wp-content/uploads/2013/09/mc_conference_day_kirsten1.pdf, accessed August 28, 2018.
70 TAA, *Minoo and Kish Islands*, Stage 1, 184.

to for said airfield.[71] The British government acknowledged that they were in no legal position to prevent Sheikh Rashid from building an airfield. A report on the potential airport at Jebel Ali (today the site of Dubai's World International Airport) recommended that the airport be built closer to town, although such a move would interfere with aviation equipment operating at Sharjah's aerodrome.[72] From what the Dubai International Airport is today, it is rather bewildering to sift through diplomatic correspondence and find that the British government saw little harm in Rashid building "a sand strip" for 56,000£.

At the same time, British officials did not want to disrupt the delicate balance of agreements they were orchestrating between Dubai and its Trucial neighbors, including Sharjah.[73] Part of this dance would be performed by diplomats who "must not discourage the Ruler of Sharjah by being seen openly to divert civil air traffic from Sharjah airport, which he would regard as an unfriendly and unhelpful gesture on our part."[74] Never one to miss an opportunity for synergy between aviation and hospitality, "The Ruler [of Dubai] . . . also asked the Company to look into the possibility of constructing a hotel near to the airport. He wishes this to have twenty single and ten double rooms, all air conditioned, with facilities for European cooking, bars, etc."[75] As a result, with the completion of the new terminal and airstrip "most, and probably all, the civil aircraft at present using Sharjah aerodrome would transfer to Dubai when the new aerodrome was available, and the proposed new hotel would no doubt cater for the passengers who now used Sharjah Fort."[76]

By 1965, the British government acquiesced to Dubai's request, integrating the airfield into a Persian Gulf consortium of airports headquartered in Bahrain. The Civil Air Attaché in Beirut was more specific and expansive: "The inclusion of Dubai into the [aviation] Plan provide a splendid opportunity for demonstrating to the Ruler our willingness to help him in putting his new Airport on the international map."[77] Sharjah's ruler protested, especially after five of the six flights

---

71 R.J. Beveridge, Minutes, "Building of an Airfield at Dubai," BA 1383/1, BNA.
72 See R. Lewis and P. Hadfield, "Report on Preliminary Survey of Aerodrome Site at Jebel Ali," in Hadfield to G.C. Lowe, Ministry of Transport & Civil Aviation, February 3, 1959, BA 1383/1, BNA.
73 G.C. Lowe MTCA to P.D. Stobart, "Proposed Airport for Dubai: encloses copy of site inspection report," March 12–13, 1959, BA 1383/1, BNA.
74 W.H. Luce to R.S. Stewart, 19 July 1965, "Civil Aviation," 1965, FO 371/179930/27, AGDA.
75 See "Minutes of Meeting held in Berkeley Square House on 30th June 1959 to discuss the proposed new airfield project at Dubai," 3, in BA 1383/6, "Dubai Airfield: Record of Meeting at M.T.C.A.," "Dubai Airfield: Record of Meeting at MTCA," BA 1383/6, BNA.
76 Ibid.
77 Civil Air Attaché to H.G. Balfour-Paul, Esq., HBM Political Agency, Trucial States, Dubai, FO 1016/837, "Dubai Airfield," BNA.

scheduled for Sharjah were reassigned to Dubai. The crown did not concur, it's agent noting, "We can see nothing but good resulting from including Dubai in the [Gulf aviation] Plan as the Ruler obviously wishes the airport to be used for international air navigation."[78] J.R. Rich relayed to the British Residency in Bahrain that "[when] the General Manager of Gulf Aviation was in Dubai on 10 May," following the announcement of the flight transfer, "he [(the Gulf Manager)] was summoned by the Ruler of Sharjah, who was in an ugly mood and asked for an explanation."[79] To register his displeasure even more vehemently, Sharjah's Sheikh suggested that the sixth flight also be reassigned to Dubai. Sharjah also agreed to sell its' new airfield equipment to Dubai. Ultimately, this row involving rival Persian Gulf airfields and Gulf Aviation was merely the first in a series of conflicts that would see Dubai pull out of the Gulf Aviation network, establish its own long-haul airline (Emirates), and strategically expand a commercially lucrative and strategically innovative airport.

At the time, however, "Gulf Aviation [was] content to route all their services to or through Dubai."[80] In the event, bridges were burned between the Sheikhs of Dubai and Sharjah. British officials took note, chalking it up to raw feelings between the Trucial States and the British crown. Rich concluded, "Shaikh Saqr's outburst against Gulf Aviation took place just before the Minister of State's visit and doubtless was a facet of his new truculence and open hostility to Britain, in which his jealousy of Dubai is also an important factor."[81]

Prior to independence, Dubai, Abu Dhabi, and Doha (Qatar) received essentially the same support for airfields from the British crown.[82] While there was little inkling that these three modest air strips would one day become the most opulent hubs in the Gulf airlines' strategies to disrupt the European-Asian commercial avia-

---

78 J.A. Taylor, Dubai Airport Manager to H.M.B. Political Agent, Dubai, "Development and Operation. Of Dubai Airport," April 21, 1965, FO 1016/837, "Dubai Airfield," BNA.

79 J.R. Rich to British Residency in Bahrain, May 10, 1965, FO 1016/837, "Dubai Airfield," BNA.

80 Ibid.

81 Ibid.

82 In terms of airport infrastructure, Abu Dhabi pioneered aesthetic terminal design. In the mid-1970s, government officials hired airport starchitect, Paul Andreu, to design a terminal that catered primarily to transit passengers, for the many reasons discussed in the text above. Based in part on Andreu's designs for Charles De Gaulle International Airport in Paris, the disc shaped structure facilitated adjacent parking for a number of wide-body jets. He would design a second terminal in the late 1990s and early 2000s on a similar architectural model. Unsurprisingly, Andreu would be called upon to help with the design of Dubai International Airport's Terminal 3 in the early 2000s. On the Abu Dhabi airport and its design for transit passengers see Paul Andreu, *J'ai Fait Beaucoup D'Aerogares* (Paris: Descartes & Cie., 1998), 27–28. Also see Philip Dodidio, *Paul Andreu: Architect*, with a Preface by Adrian Frutiger (Basel, Switzerland: Birkhauser, 2004), 54–61.

tion market, Dubai did give some cause for concern among its fellow city-states. To the consternation of British officials, their Trucial counterparts, as well as Gulf Aviation, Dubai's leaders refused to walk in lock step with the other emirates as far as commercial aviation policy was concerned.[83] Thus, instead of working through Gulf Aviation to attract carriers and develop route networks for the benefit of all, Dubai insisted on an "open skies" policy, which undermined Gulf Aviation generally and the prospects of the airports of fellow emirates particularly. Bewildered British officials commented: "If . . . Dubai is only interested in giving away traffic rights at its own airport without getting any aviation concessions outside its own territory in exchange, there seems to be no impediment in international law to its doing so on the basis envisaged . . . though it would be very unorthodox."[84] Widebody jets arrived in the 1970s, as well as a runway which "allow[ed] for the eventual handling of Boeing 747 and Concorde aircraft."[85]

Emirati leaders in Dubai had their sights set on bigger targets than Sharjah's modest transit traffic. Ultimately, it intended to challenge Bahrain for Persian Gulf entrepot preeminence. Bahrain and Gulf Aviation officials saw the writing on the wall and scrambled to implement a hubbing strategy. British officials noted that "plans were going ahead for the official opening of the new Gulf Hotel in Bahrain on September 1969," a facility that would be "run by Gulf Aviation under a management contract." The trouble was that British Overseas Aviation Corporation had made a similar proposal to Dubai and "plans were also going ahead for new hotels under similar arrangements in Abu Dhabi and Doha, and an initial agreement had been made over the latter."[86]

Hotel openings were harmless enough, but the spirit of shaking up Persian Gulf aviation was alive and well in Dubai. While Bahrain made slow headway with British officials in securing upgrades for its own airport, Dubai pressed ahead. "As you know the Lower Gulf is already over-served by modern international airports and international operators will not be persuaded to continue to use Bahrain merely because its airport offers similar facilities to those at Doha, Dubai, and Abu Dhabi," British officials observed.[87] Internally, there was no sense

---

83 Douglas to Immediate Bahrain Residency, "UAE Independence: Civil Aviation," AVIA 120/25, "The Future of Civil Aviation in the Persian Gulf," Part D, BNA.

84 Ibid, pp. 2–3.

85 Oxford Economics, "Explaining Dubai's Aviation Model," June 2011, A Report for Emirates and Dubai Airport, https://www.oxfordeconomics.com/my-oxford/projects/128910, accessed January 24, 2022.

86 "Call by Mr. Orpin of BOAC ," in "Visit of the Ruler of Dubai and Aviation Law in the Persian Gulf," FCO 14/591, AGDA.

87 Hugh Perry, July 30, 1968, "Restricted," 23224/7/68, "Bahrain Airport," 1968, FCO 8/562, AGDA.

that such efforts would challenge the British, one functionary writing: "These airports may represent a wasteful duplication of expensive facilities but for reasons more related to local pride than economics they have been built and will therefore be utilised." With a bit of humility, the same official hedged: "with finance no problem they are likely to be further improved as time goes on so that they too can take any number of large aircraft."[88]

An aviation attaché detected Dubai's ambitions in November 1968. His concerns about Dubai's threat to Bahrain deserve to be quoted at length because they foreshadow its later leadership in the region. The attaché noted:

> A major preoccupation in Dubai is the expansion of their facilities so as to attract to Dubai airlines at present transiting the area with stops at, primarily, Bahrain. The authorities have recognised that a very large part of the Gulf traffic is transit traffic, only stopping in the Gulf because of the technical need to refuel at an intermediate point on a trunk route; they therefore see this traffic stopping at the airport which offers the best facilities, in their widest sense, to the airline operators. [Dubai's] 4m*f* expansion, largely being spent on a new terminal facility allegedly designed for the 747 age has this kind of traffic primarily in view. There is no feeling that the present facilities are inadequate for the traffic which flows in and out of Dubai, but it is recognised that business would improve and additional revenues would be earned if the trunk operators could be tempted to go to Dubai instead of, as at present, Bahrain. This kind of competition is likely to generate progressively more disrupting forces, and some consideration needs to be given as to how it can be contained.[89]

"Disrupting forces," indeed. As would later happen in Singapore, Dubai's top brass "thought big" in planning its hubbing beachhead. Whatever the city-state's reconnaissance methods, they anticipated the next step in civil aviation development. One year after this remarkable analysis of Dubai's airport expansion, British officials noted that "Qantas now [intends] to introduce Boeing 747s on their schedules through Bahrain in November 1970; BOAC followed suit early in 1971." These observations were made in the context of Bahrain's scramble to accommodate the quantum jump in aircraft technology. Meanwhile, in a prosaic afterthought that reflected what had already taken place farther south, "Mr. Wilson . . . thought it possible Air-India might, at some stage, operate 747s through Dubai."[90]

Too often, post-independence British archival files carry a whiff of hubris. As a result, imperial officials sometimes badly misjudged—and even dismissed—what they imagined to be the unthinkable: that one of the lone emirates might start its

---

88 "Bahrain Airport," FCO 8/562/36, AGDA.

89 "Report on Visit by Civil Air Attaché to Persian Gulf, November 6–17, 1968," "Visit of the Ruler of Dubai and Aviation Law in the Persian Gulf," FCO 14/590, AGDA.

90 "Call by Mr. Orpin of BOAC," "Visit of the Ruler of Dubai and Aviation law in the Persian Gulf," FCO 14/591, AGDA.

own airline. The Civil Air Attaché raised this hypothetical during his November 1968 visit to the Persian Gulf. Almost as a glib throw-away, he included an attachment to his report, entitled, "Starting an Airline in the Persian Gulf – and Why Not!" He dismissed out of hand the recurring reasons why such rumors had been floated: indigenous nationalism, inadequate services, or empresarios looking "to make a fast buck." It was incumbent on British officials that should they encounter such schemes, to "discourage enterprises of this sort since it is both in the British interest and in the interests of the indigenous rulers that this should be done." Anticipating certain windfalls from the post-independence shakeout, he continued, "we have a financial stake in Gulf Aviation, which at present makes a profit because of its exclusive right to . . . traffic between the states for whose aviation affairs we are responsible . . . which supports BOAC's operations through the Gulf, and whose demise would leave a vacuum that would be filled by, and thus give strength to, competing airlines in the region." He went on to express his "hope" that Dubai would sign on to the ownership and management of Gulf Aviation, since, in his humble opinion, "The setting up of a rival airline in any one of the Gulf states would be expensive, and a guaranteed money loser for them and for us." Even more improbable, he concluded, would be such an operation that would function internationally, since it is "manifestly impossible for any airline to pay its way operating wholly within any one of the sheikdoms."[91]

Three years later, what once had been a hypothetical presented itself as a very real path forward that Dubai's sheikh might pursue after independence. J.T.T. Boulton surmised that Sheikh Rashid's real motivation might be "to try to operate a Dubai national airline by a restrictive traffic rights policy." Like the civil air attaché, this official's hubris momentarily quenched the feasibility of such an endeavor. Boulton flat underestimated the city-state's ability to run an efficient, rationally executed—not to mention profitable—business plan, writing "These days too many [national] airlines are prestige outfits set up primarily to allow the Head of State or other dignitaries to travel abroad in an airline bearing the national flag." Further dismissing the Emirati's ability to do its own bidding, he continued, "We could not guarantee that it would get any traffic rights to London, and, to put it mildly, we could not undertake to pursue its interests vigorously with foreign governments." His final analysis favored the status quo: "Experience in the Middle East generally suggests that a Dubai national airline would find it so difficult to make a living that it would find itself having to indulge in every crooked

---

91 "Report on Visit by Civil Air Attaché to Persian Gulf, November 6–17, 1968," in "Visit of the Ruler of Dubai and Aviation law in the Persian Gulf," FCO 14/590, AGDA.

game in the business."[92] Today, Emirates boasts one of the highest frequencies of wide-body arrivals and departures at London Heathrow, effectively carrying a lion's share of European-based and Europe-bound passengers to and from Southeast Asia and Australia.

***

Without ready access to Iranian Archives, the historian must look elsewhere to learn the fate of the Taliesin plan for an aviation and entertainment entrepot on Kish and Minoo Islands.[93] The Ministry of the Interior's request for a modified development plan in 1969 suggests that the government was still keen on the program, but only on a more modest level. The Shah initiated limited development on Kish Island, which he anointed his own island playground.[94] The *New York Times* estimated that the resort would require $100,000,000 for completion (likely an exaggerated amount)—yet another sign of the decadence.[95] At Kish, the Shah met with prominent officials, including Nelson Rockefeller in 1976.[96] He had a hotel and spacious guest facilities built for his guests. French chefs swooped in to cater to discerning palates. Lou Martin, a sometime pilot for the Shah's family, wrote of his experience there during a visit in the late 1970s. After flying the

---

92 J.T.T. Boulton, "Civil Aviation in the Gulf," January 6, 1970, "Air Services Agreement: Netherlands Airline KLM Access to Dubai," FCO 14/816, AGDA.

93 Amery, the Iranian architect noted, "The Minoo Project would have been one of the many tourist attractions in the world which have hotels, restaurants, etc., but our plan would have been entirely different. In our feasibility studies we proposed that Minoo should be developed because it had a lot of infrastructure already there, such as the Abadan Airport close-by, also electricity and water. I thought it better to develop Minoo which would not require so much extra budget but Mr. Alam [the Ministry of Interior] voted to do the development on Kish Island instead." Of the Shah's reaction to the proposal for Kish, Amery noted, "When I showed the preliminary drawings to the Shah, he was very enthusiastic and liked the project but unfortunately the political situation was not favorable and the project did not materialize." (e-mail from Nezam Amery to the author, December 18, 2007).

94 Improbably, the Shah alienated the economic elite of Iran by insisting that they invest in his tourist enclave at Kish during the 1970s. Cynthia, wife of American ambassador to Iran, Richard Helms, wrote in her memoir: "The wealthy businessmen were often exploited. For example, they might be obliged to purchase expensive villas on the resort island of Kish." See Helms, *An Ambassador's Wife in Iran* (New York: Dodd, Mead & Company, 1981), 175. Helms' memoir also provides one of the only images of the Shah's exclusive retreat and it appears to have been built after the manner of the Peters and Amery design for Kish and Minoo (82).

95 A.M. Rosenthal reported that the Shah was building the enclave "that he hopes will get him a little more oil money – from Persian Gulf sheiks and other Arabs who now go to Beirut for a good time'. See "Shah, Confident of His New Power, Sells Iranian Oil to Whom He Pleases," *New York Times*, March 31, 1974, *Pro-Quest Newspaper Database.*

96 Amy Fitch to Evan R. Ward, e-mail, August 20, 2007. *The New York Times* also corroborated this meeting in 'Rockefeller Talks with Shah", March 25, 1976, 44, *Pro-Quest Newspaper Database.*

Shah's brother to the island, a waiting limousine whisked Martin away to a bespoke guesthouse with appurtenances including a wet bar and French cuisine "equal to the finest restaurants in Paris."[97] He later couched his experience in the political context of the times, writing:

> I spent the next two days strolling on the white sandy beach, shopping in the bazaar, indulging in great food and drink, and just plain relaxing. . . . This opulent no expense-sparred trip took place as Iranians, denied their fair share of the oil wealth, were assembling in Tehran and other major cities to march in violent protest against Prince Pahlavi's oldest brother, the Shah.[98]

Curiously, the fundamentalist Khomeini regime, which overthrew the Shah's government in 1979, integrated the free port concept of Kish Island into successive development programs. Today it is a poor brother to its neighboring emirate—Dubai—some hundred miles across the Persian Gulf (serviced no less from the Dubai airport by Kish Airlines—founded in 1990). Many Iranians fled to Dubai in the wake of the Shah's regime, not to mention during the rise of fundamentalist governments. Abadan and adjacent Minoo suffered a much more devastating fate. Its urban fabric was wiped clean during the Iran-Iraq War—the target of disputes over access to the Persian Gulf. The population of the city plummeted towards oblivion. Today, the city is rarely cited as a tourist destination even though its renaissance has restored some measure of strategic importance to the region.

To the chagrin of the British and neighboring emirates, Dubai's "open skies" stance extended to its strategy for launching an airline in 1985. The very name of the carrier—Emirates—appropriated the identity of a multi-nodal state, as well as the color scheme of the flag, for its livery. Prior to this point, Gulf Air (previously Gulf Aviation) juggled a variety of schedules to serve as *de facto* national carrier. Abu Dhabi, Qatar, Oman, and Bahrain pooled financial control of the airline, but still offered service to Dubai. Dubai's move to start its own airline triggered fateful repercussions. Gulf Air eventually reduced flights to the Dubai, prompting Emirates to target overseas destinations, a move that fed its ambition as a global carrier. Gulf Air's efforts to exact landing fees from legacy carriers who serviced Dubai, with its "open skies" policy—including Singapore Airlines, KLM, and Cathay Pacific, simultaneously alienated Dubai from civil aviation equipoise with its neighbors. Subsequent collaboration with Pakistan International

---

97 Lou Martin, *Wings over Persia*, Fifth Edition, (Victoria, British Columbia: Trafford, 2005), 165.
98 Ibid., 166.

Airlines opened doors to the Indian subcontinent whilst also shutting the door on further expansion by Gulf Air into lucrative routes Gulf-Pakistani routes.[99]

Farther east, leaders in the city-state of Singapore made their own plans for tourism development based on the centrality of their location at the mouth of the Malacca Straits. The presence of world-class infrastructure provided the basis for a unique tourism development strategy, the rise of a globally preeminent airline, and accommodations to coddle weary travelers hopping around Southeast Asia.

---

99 These intrigues are discussed at length in a pair of detailed journalistic accounts of Emirates early history. See Angela Dixon, "Survey of the United Arab Emirates: Airline Competition [Heats] Up," *Financial Times*, January 21, 1986, Factiva. Also see Kathy Evans, "World Trade News: Feather Ruffled as New Airline Takes to Gulf Skies," *Financial Times*, October 23, 1985, Factiva.

Chapter 2
"An Entrepot for Tourists:" Gateway Travel in the Indo-Pacific, 1934–2019

> I developed the theory that . . . we could develop Singapore as a sort of entrepot for [tourists]. We don't have enough interesting things for tourism . . . But what we [have] is a runway, our airport, which will receive any airplane, and see to it that we always keep a runway which can receive the largest plane which need the longest runway.
> – Albert Winsemius, Economic Consultant to the Singaporean Government (1982)[1]

> Transit passengers have choices. They do not have to come here. We do our best to make it worthwhile for them to transit through Singapore, but other airports are working hard too.
> – Minister Khaw Boon Wan, Opening of Terminal 4 (2018)[2]

> I can't believe this airport has a butterfly garden and a movie theater. JFK is just salmonella and despair.
> – Rachel Chu, *Crazy Rich Asians* (2018)

Introduction

In the spring of 1959, Singapore's Minister of Finance, Dr. Goh Keng Swee, decided against renewing the contract of R.A. Howlett, a tourism specialist employed by the Singaporean and the Malayan governments. Howlett had spent the previous two years creating a promotional campaign to attract North American and European travelers to Singapore. What Singapore lacked in tourist attractions, including mountain resorts and an extensive coastline, might justify Malaya's retention of Howlett on a full-time basis, only not funded with Singaporean tax-payer's money.[3]

---

1 Albert Winsemius, Oral History, Tan Kay Chee, interviewer, National Archives of Singapore (cited hereafter as "NAS"), Accession Number 000246, "Economic Development of Singapore," August 31, 1982, reel 12 of 18, 143–144.
2 Khaw Boon Wan, "Speech by Minister Khaw Boon Wan at the Official Opening of Terminal 4, Changi Airport," August 3, 2018, https://www.mot.gov.sg/news-centre/news/Detail/speech-by-minister-khaw-boon-wan-at-the-official-opening-of-terminal-4-changi-airport, accessed June 16, 2021.
3 Even today, Singaporean scholars have had difficulty distinguishing between the island's infrastructure and its relationship to tourism there. See, for, example, the "tourism" entry for the 2006 *Singapore: The Encyclopedia*: "Tourism: Given its small size, Singapore is not a natural tourist destination, but tourism has developed into a major industry due to the city-state's prime location as a gateway to Southeast Asia. It has achieved this through its excellent communication links and world-class Changi Airport," see Tommy Koh, ed. in chief, *Singapore the Encyclopedia*, (Singapore: Editions Didier Millet and the National Heritage Board, 2006), 568.

https://doi.org/10.1515/9783111326641-003

Goh's assessment accurately measured the pulse of the People's Action Party (cited hereafter as "PAP"), architects of a city-state in the opening stages of economic formation. Goh, who would be tasked by the iconic father-figure of Singaporean history, Lee Kwan Yew, with creating an export-led, politically stable island nation, believed there were higher economic priorities than tourism, such as industry. Though Goh circled back as a promoter of Singaporean tourism, which included spending money on airports, an airline, and attractions that would benefit Singaporeans as much as they did tourists, he initially eschewed conventional methods of promoting tourism development championed by multilateral aid agencies and the privately funded Pacific Area Travel Association (i.e., expensive foreign promotional campaigns and an almost exclusive emphasis on luxury hotel construction).[4] What's more, Singapore could profit already from the most abundant group of travelers arriving on the island: transit passengers.

This chapter is not a history of Changi International Airport, Singapore Airlines (cited hereafter as "SIA"), nor of Singaporean tourism development, per se. Instead, it is the study of a unique institutional approach to tourism development embodied in Goh Keng Swee's reaction against prevailing methods of tourism development promoted by Western modernisers in the mid-twentieth century. Singapore prioritised spending on airports and an airline to service its own fledgling nation state, as well as profiting from transit passengers traveling throughout Southeast Asia. While it would be too simplistic to posit one administrator's response as the origin of an over-arching, almost seventy-year long history of tourism policy, Goh's reaction epitomised the monetarist, state-development mindset of Singapore's founders.[5] Ultimately, the idea of an "entrepot for tourists" evolved over time, from the death throes of British colonialism through construction of Changi International Airport as a destination in its own right.

Temporally, this chapter covers three major eras, highlighted by the 1981 opening of Changi International Airport, and culminating with the construction of the Jewel Project in 2019.[6] The first period (1937–1955) encompasses British-led deci-

---

4 Can Seng Ooi concurs with the Singaporean predilection to design tourist attractions that also engage locals. See "State-Civil Society Relations and Tourism: Singaporeanizing Tourists, Touristifying Singapore," *Sojourn*, 20:2 (2005), 249–272.

5 Katalin Volgyi stresses the PAP's preference for state owned industries in "A Successful Model of State Capitalism: Singapore" in Miklos Szanyi, ed., *Seeking the Best Master: State Ownership in the Varieties of Capitalism* (Central European Press, 2019), 276–296.

6 In terms of tourism history, Alfredo Mena Navarro, Fernando Almeida Garcia, and Rafael Cortes Macias, identified the 1960s and 1970s as the "take-off phase," which corresponds roughly to the "before Changi" and "after Changi" model I have proposed here; a "development phase" during the 1980s; and, a "maturity phase," running from the 1990s to 2015. See "Evolution of Singapore Tourist Policy (1965–2015)," *Cuadernos de Turismo*, number 41 (2018), 703–706.

sions to favor military over commercial aviation at Changi Airfield as well as limited plans to promote Singapore's growth as an aviation entrepot at the new Paya Lebar Airfield, completed in 1955. During a second period (1958–1980), Singaporean officials devised an aviation-centered, transit-passenger based tourism strategy, started Singapore Airlines, and planned to build Changi International Airport. Finally, after 1981 state leaders sought for global recognition through its airline and airport. Throughout these stages Singapore's aviation infrastructure evolved from a simple entrepot for air passengers flying to other Asian cities, to a destination, with entertainment, shopping, and dining lauded by transit passengers.

Said another way, in the immediate aftermath of World War II, British officials intended to maintain the status quo in Singapore's aviation infrastructure. They restricted military air movements at Changi Airfield and limited expenditures on civil airports at Kallang and Paya Lebar Airfields. Next, during the 1960s and 1970s, PAP bureaucrats emphasised Singapore's role as an entrepot offering dependable, efficient, and comfortable infrastructure (airport) and superstructures (hotels and airport amenities) to transit tourists fanning out across Southeast Asia (preferably on SIA). This period of consolidation mirrored a general climate of state-directed investment not only in manufacturing, but also services, including the transformation of Singapore Airlines into an almost exclusively long-haul airline. Finally, aviation infrastructure entered a new phase in the 1980s with the opening of Changi International in 1981. It was at this point that state planners pursued not solely preeminence among not only East Asian airports, but global acclaim from international travelers. Bureaucrats carefully tracked Singapore's expanding reach by annually recording new destinations serviced by inbound and outbound Changi flights. They also paid close attention to the airport's global status in customer satisfaction surveys charting the "world's best airports." With the recent creation of Changi's Jewel, the airport infrastructure became the spectacle, a trend that will only gain traction as hubs in East Asia and the Middle East vie for global hub supremacy.

More broadly, this chapter also explores the origins and trajectories of competing air hubs within the Indo-Pacific region vis-à-vis Singapore's evolution. Hong Kong is the closest point of comparison. Direct competition for the Kangaroo Route began after World War II when Cathay Pacific, Qantas, and British Overseas Airways Corporations competed on the Southeast Asia to Europe routes through Hong Kong. Next, the Indian sub-continent was the great mystery of Indo-Pacific commercial aviation development, where neither Mumbai (a former stop on ocean going voyages between the Suez Canal and Asia) nor the new capital, Delhi, engineered a plum aviation entrepot at the basin's terrestrial anchor. Most of its long-haul travel flowed through Singapore, Hong Kong, or later, Dubai. Finally, the Persian Gulf, as the previous chapter illustrated, always had a role in the traditional route between Southeast Asia and Europe, what with its strategic position as a refueling center. Dubai supplanted stopover points in Sharjah (twenty miles away from the opulent Emirate) and Bahrain, coming into its own starting in the 1980s. Dubai pioneered the outsize duty-free experience, competing directly for passengers choosing between dueling hubs across the breadth of the Indian Ocean.

Theoretically, this chapter also builds on the work of Arang Keshavarzian and Waleed Hazbun, who argue that emerging nations in the decolonised Middle East harnessed transportation infrastructure to challenge networks established by their Western counterparts. This approach has been applied here in a Southeast Asian context.[7] These networks, in both their Middle Eastern and Southeast Asian contexts respectively, served as alternatives to enterprises and practices originating in Europe and the United States. More specifically, this chapter builds on a growing body of work that identifies East and Southeast Asian imaginaries in the South Pacific and Indian Ocean basins. Engseng Ho's study of past and present "inter-Asian" networks emphasises that "while many Asian societies today have indeed been shaped by the modern West and reflect it . . . there were and are many mobile societies that sustain relations across Asia beyond and before globalization's reach."[8] More recently, Jonathan Bollen's innovative work on aviation and theatre networks in the South Pacific and East Asia links entertainment to mobility. Bollen posits a growing number of historians whose work "run[s] counter to conventional European-American narratives, challenging both the legacy of the British Empire in Asia and the discourse on 'Americanisation' across the Pacific."[9] These theoretical disruptors of western-anchored globalisation extend even to Changi International itself, as Rachel Bok demonstrates in her article, "Airports on the Move? The Policy Mobilities of Singapore's Changi Airport at Home and Abroad." In emphasising the national symbolism of Changi International's and Singapore Airlines' enterprises, Bok argues, institutional actors transcended the purely economic significance of these entities. In the broader context, then, "here is the pressing need to broaden the empirical horizons of . . . mobility scholarship from almost exclusively western contexts, in order to acknowledge diverse policymaking motivations apart from neoliberalising ones." "In this regard," she continues, "it is worthwhile turning our

---

7 See Keshavarzian, "Geopolitics and the Genealogy of Free Trade Zones in the Persian Gulf," *Geopolitics*; Hazbun, *Beaches, Ruins, Resorts*, 216–220.
8 Engseng Ho, "Inter-Asian Concepts for Mobile Societies," *The Journal of Asian Studies*, volume 76, number 4 (November 2017), 908.
9 Jonathan Bollen, *Touring Variety in the Asia Pacific Region, 1946–1975* (Transnational Theatre Histories) (Palgrave Macmillan, 2020), 5.

attention to nonwestern settings, where neoliberalism's utility as a principal, coherent analytic is far less robust."[10]

In terms of its disciplinary significance, this chapter complements Joan Henderson's decades of work on Singaporean tourism development, marketing, and comparative positioning from a purely historical and archival perspective.[11] The same can be said for T.C. Chang's work on Singapore's regional positioning as a center for tourism (as a complement to that on offer from neighboring states), though the claim made here is that Singapore's leaders set their sights on regional prestige in the field of East Asian transit options and global preeminence among its peer airports.[12] Finally, it temporally extends the story told in Jonathan Bollen's work. He establishes that following World War II, airline executives pegged the soon-to-be city-state as an important node in regional travel (not to mention entertainment as well). As a result, Bollen notes, "As the nation-state of Singapore emerged from British colonial rule and separated from the Malaysian Federation to become an independent Republic, the integration of tourism and trade in entertainment became a national concern."[13]

The idea of "an entrepot for tourism" contributes to the idea that Singapore's role as a Southeast Asian intermediary preceded independence and anticipated its pivotal economic and geo-political space in the Indo-Pacific. Singapore had long served as a trading nexus, even before the arrival of the Portuguese, Dutch and British. Indeed, the late twentieth century discovery of Chinese porcelain dating to the fourteenth century corroborated this claim. These developments led to a revision of Singapore's role as a regional entrepot that predated the arrival of Sir Stamford Raffles in 1819, and resonate with the nation-state's broader global

---

10 See Rachel Bok, "Airport on the Move? The Policy Mobilities of Singapore Changi Airport at Home and Abroad," *Urban Studies*, 52:14 (November 2015), 2729.

11 Joan C. Henderson, "Destination Development and Transformation: 50 years of Tourism After Independence in Singapore," in *International Journal of Tourism Cities*, 1: 5, 269–281. Henderson has covered the Singapore's tourism branding of Singapore in several publications including "Case Study: Uniquely Singapore? A Case Study in Destination Branding," *Journal of Vacation Marketing*, 13: 3 (2007), 261–274; on tourism development generally, see Henderson, "Destination Development," *Journal of Travel & Tourism Marketing*, volume 20, number 4, 33–45. See also Henderson, "Tourism and Development in Singapore," chapter 8, in Eduardo Fayos-Sola, ed., *Tourism as an Instrument for Development: A Theoretical and Practical Study* (Emerald Books, 2014), 84–88.

12 T.C. Chang has addressed Singaporean tourism development in a regional context in "Regionalism and Tourism: Exploring Integral Links in Singapore," *Asia Pacific Viewpoint*, 39:1 (April 1998), 73–94.

13 Bollen, 86.

ambitions.[14] Further, these discoveries called for a reassessment of the almost exclusive association between British colonialism and Singapore's post-colonial trajectory.[15]

In the context of decolonisation, Lee Kwan Yew, Goh Keng Swee, and a host of technocrats chronicled in this study, built on the British foundation of Singapore as a regionally prominent military and commercial air hub, but transcended this model to create a center of *global* influence.[16] While they took advantage of the benefits of colonisation left in the wake of the British departure, including the very runway on which the Changi Airport was constructed, Singaporean politicians and experts used alternative strategies to multilateral aid agency prescriptions in launching their own aviation-led strategy targeting transit passengers, all the more striking in the case of Changi since it was proposed and built at the very moment of economic retrenchment in Europe and the United States.[17]

✳✳✳

---

14 This corresponds with the careful efforts of Chong Guan Kwa, Derek Heng, Peter Breschberg and Tan Tai Yong's to write a "history from as Asian perspective, rather than a European one that makes Singapore's past a part of British colonial history." See Kwa, et al., *Seven Hundred Years: A History of Singapore* (Singapore: National Library Board, 2019), vii. Historian John Curtis Perry also situates Singapore's extended history (back to the 13th century) within the study of global city states. See John Curtis Perry, *Singapore: Unlikely Power* (New York City: Oxford University Press, 2017).

15 Constance Turnbull's *A History of Modern Singapore, 1819–2005* (Singapore: National University of Singapore, 2020), dominated the historiography for much of the twentieth century, framing the city-state's development to its British roots.

16 While I would not characterise this study as "revisionist," it does support Michael D. Barr's assertion that the *longue durée* of Singaporean history should be viewed within a broader than colonial context. See Barr, *Singapore: A Modern History* (London: Bloomsbury, 2019), 10.

17 Barr's interpretation of post-colonial Singaporean history attributes many of the benefits (and subsequent economic successes) accrued from institutions and practices left in place by the British when they left in the late 1950s. See *Singapore: A Modern History*, xiii–xiv. With reference especially to the location of Changi International, see Loh Kah Seng, "The British Military Withdrawal from Singapore and the Anatomy of a Catalyst," in Derek Heng and Syed Muhd Khairudin Aljunied, *Singapore in Global History* (Amsterdam: Amsterdam University Press, 2011). Significant emphasis in this study centers on Goh Keng Swee, who has been the subject of several chapters in the recently re-issued *Lee's Lieutenants: Singapore's Old Guard*, edited by Kevin Yi Tan and Lam Peng Er, revised edition (Singapore: Straits Times Press, 2018), including Tilak Doshi and Peter Coclanis' essay, "The Economic Architect: Goh Keng Swee," 80–109.

Colonial Entrepot: 1937–1955

Singapore emerged as an aviation entrepot during the early twentieth century. In 1937, one hundred and eighteen years after Sir Stanford Raffles claimed Singapore as a British outpost, the British crown inaugurated Kallang Aerodrome on the shores of the island-colony. A year later, in 1938, Qantas introduced flying boat service from Sydney to Singapore, with British Overseas Airways Corporation carrying passengers the rest of the way to London, on an itinerary that would later be known as the "Kangaroo Route." Originally, Qantas initiated service from Brisbane to London via Singapore in 1934, meeting up in the Lion City with Imperial Airways, which carried passengers the rest of the way to the British capital. When Qantas discontinued flying boat service in 1939, it had carried 4900 passengers between the two cities.[18] By the post-war years, Qantas was one of many national flag-carriers (the Australian government nationalised its flagship airline in 1947) calling at Singapore.[19] So well-known was Qantas's relationship with the city-state in the new age of "transit passengers" that it literally became part of the famed Raffles Hotel. As historian Gretchen Liu has written, "In the days of short-haul flights, passengers travelling, say, the Qantas-BOAC Kangaroo route to London, were obliged to night-stop in Singapore. Qantas opened their Singapore office on the ground floor of Raffles' Bras Basah wing and rented a fixed number of rooms there at a fixed rate. This portion of the hotel was soon known as the Qantas wing."[20]

No less a judge than Amelia Earhart gave the most flattering description of the airport which she first saw after flying west from Yangon to Bangkok, before turning south to Singapore in April of 1938. "The vast city lies on an island," she wrote, "the broad expanses of its famous harbour filled, as I saw them from aloft that afternoon, with little waterbugs, ships of all kinds from every port. Below us, an aviation miracle of the East, lay the new airport, the peer of any in the world."[21] The intrepid aviator's blurred distinctions between an entrepot of the sea and of the air might be easily forgiven at this early stage in commercial aviation. Indeed, at the opening of the facility, which had been reclaimed from "an evil-smelling tidal swamp," the

---

18 John Gunn, *The Defeat of Distance, Qantas, 1919–1939* (St. Lucia, Queensland: Queensland University Press, 1985), 336.

19 I'm grateful to Qantas Airlines Archival Curator David Crotty for elucidating the twisting story of Qantas's Kangaroo Route from its inception to present-day configuration by email (June 6, 2021). He is the author of *Qantas the Empire Flying Boat* (Stamford Lincolnshire, UK: Key Publishing, 2021).

20 Gretchen Liu, *Raffles Hotel* (Singapore: Landmark Books, 1992), 143–144.

21 See Amelia Earhart, "Singapore Airport: An Aviation Miracle of the East," *Brisbane Courier-Mail*, April 11, 1938, 4.

Straits Settlement Governor, Sir Shenton Thomas, likewise boasted, "What Port Said is to shipping, Singapore becomes to air traffic." Continuing he noted, "This airport is not only a key position for Empire [Airways] routes, but also the most important centre, for airlines in the Far East."[22] Indeed, Earhart's assessment was not inflated. At a key strategic point, the British had laid out an airport that in the mind of its construction chief, R.W. Caulfield, "bore comparison with any he had seen," capable of receiving any aircraft, as well as accoutered with passenger amenities like long, running canopies unfurled at a moment's notice to shield passengers from frequent downpours.[23] Ultimately, what would be stated as advantages of Changi some five decades on might have echoed the words of an envious Melbourner, who gushed about a "magnificent centre [which] contained ample buildings and accommodations for passengers and for officers of aircraft companies."[24]

Nevertheless, after World War II, Britain dictated the strategic and economic priorities of Singapore's aviation needs. Imperial priorities favored British objectives over those of the colony.[25] As a result, British planners nixed a proposal to site commercial flights at Changi Airfield, which would have required significant expense. Located on Singapore's southeastern fringe, the Changi district began as a permanent military garrison for British imperial forces prior to World War II. Following its stealth attack on Singapore, Japanese authorities turned the British cantonment into an airfield, which, after World War II, the British Royal Air Force, fortified for use by large aircraft.[26] As they had for shipping traffic during the nineteenth century, the British conditioned the island for regional aviation preeminence, noting, "Singapore, by virtue of its geographical position is a focal point for all air services [civilian and military] in South East Asia. It must inevitably be of major importance for any movements of aircraft either east from the Indian Ocean or west from the Pacific."[27]

Despite this favorable assessment of Singapore as a military entrepot, British officials debated Changi's capacity to serve both civilian and military air move-

---

22 *Wagga Wagga Daily Advertiser*, "Singapore Airport," June 25, 1937, 6.
23 *Brisbane Courier-Mail*, July 2, 1937, 21.
24 *Melbourne Age*, "Singapore Airport: Example for Melbourne," April 28, 1938, 8.
25 In this sense, a "Western" network has little to do with culture production or performance, but denotes a decision made in the interest of a European or American government or corporation versus that of a former colony charting its own economic independence.
26 "Changi Airfield," September 5, 1946, Accession Number AIR 20/7153, "Changi airfield: Development policy. Secret," Source: National Archives (Great Britain), Fonds: The National Archives (United Kingdom), Microfilm Number ARB 1769, Page/Frame 97, NAS.
27 "Air Staff Requirements for a Heavy Bomber airfield in Singapore Island, November 1946," Accession Number AIR 20/7153, "Changi airfield: Development policy. Secret," Source: National Archives (Great Britain), Fonds: The National Archives (United Kingdom) Microfilm Number ARB 1769, Page/Frame 73, NAS.

ments alike. The real question was whether a second runway, which would require either a tight fit with the existing runway or expensive land reclamation work (which the British Empire was loathe to fund) would be built. Confidential assessments pointed to the need for aesthetic considerations in accommodating civilian aircraft. "We very much hope," military officials opined, "that the Air Ministry will be able to agree that the land at present earmarked for a second runway will not in fact be required, so that a little more elbow room can be made available for the development on the civil side of what we hope will be a very effective British shop-window in the Far East."[28] Officials crowed that it would be no less a marvel than Kallang, "capable of taking the largest aircraft in the world."[29] In the end, however, due to the costs of reclaiming land for a second runway, British officials maintained the status quo, with Changi as a military airfield and Kallang Aerodrome as the primary commercial aviation venue.[30]

Kallang Aerodrome adequately serviced regional and limited intercontinental routes through Singapore from the post-war years through the mid-1950s. Capacity reached, British authorities built a modern airfield at Paya Lebar, located some five miles from downtown Singapore. Malayan Airlines, an upstart with British origins as a shipping company, serviced flights to Bangkok, Saigon, and Hong Kong. As noted above, early iterations of the Kangaroo Route, linking Australia to Asia and points north to Europe, also called at Paya Lebar. Emblematically, the closure of Kallang Aerodrome in the mid-1950s gave way to offices for the PAP, the political force which would engineer Singapore's early efforts at travel and tourism.[31]

The British made plans for the larger airport at Paya Lebar beginning in 1951.[32] Local politicians complained that the British had moved too slowly, "frustrating Singapore's development into the air-crossroads of the Far East."[33] It fi-

---

28 J.B. Jonston to T.C.G. James, Esquire, October 22, 1948, Accession Number AIR 20/7153, "Changi airfield: Development policy. Secret," Source: National Archives (Great Britain), Fonds: The National Archives (United Kingdom) Microfilm Number ARB 1769, Page/Frame 93, NAS.

29 American Associated Press, "Changi will be made a modern airport," *Melbourne Argus*, June 16, 1947. For a more extensive account of the immense expense of human labor invested in rehabilitating Changi after World War II as "the British air-centre of South-east Asia," see *Cairns Post*, "Singapore Facilities to be Enlarged: Expand Changi Airport," July 8, 1948, 1.

30 *Burnie (Tasmania) Advocate*, "Changi Project Bypassed," January 1, 1949, 18.

31 Koh, ed., "Kallang Airfield," *Singapore: The Encyclopedia*, 272.

32 According to the *Melbourne Age*, the "Kangaroo Route" grew out of the mail delivery service between Australia and Sri Lanka during World War II. See "Origin of the Kangaroo Service," September 27, 1952, 15.

33 American Associated Press, "Singapore Hurries New Airport," *Melbourne Herald*, August 20, 1953, 18.

nally opened in August of 1955, with a "7500 ft. runway . . . able to take the Comet, the world's first jet liner."[34] Paya Lebar more than doubled passenger capacity at Singapore, from 143,000 passengers in 1954, to up to 300,000 at the new airport. Capacity was not the only added benefit of Singapore's new airport, which would handle over 1.7 million passengers by 1970. As the Colony of Singapore *Annual Report* for 1955 observed, "When the planned equipment of radio navigational aids has been completed, and surveillance radar has been added, the runway should be usable in almost any weather conditions, by the largest civil aircraft now flying or likely to fly for many years to come."[35] By 1959, global capacities were in place, and, as officials noted, "Singapore can . . . justly be proud of her International Airport with a runway capable of handling any aircraft now flying in the world, and with all modern aeronautical facilities available to protect air navigation in this part of the globe."[36] The new runway accommodated jet aircraft, as well as regular service by the Concorde.[37]

In the broader context of Britain's Asian presence, Singapore kept pace with neighboring entrepot, Hong Kong. While the latter's Kai Tak airfield commenced service in 1925, it would not be until 1936 that Imperial Airways initiated service between the Pacific outpost and London, via Penang.[38] The more immediate advantage of Hong Kong lay in the fact that a private shipping and sugar company, Butterfield and Swire, launched a "hometown" airline in 1946, Cathay Pacific Airways. As Adrian Swire, a descendent of the company's founder John Swire, noted in an interview some years later, Hong Kong's geography was a principal reason the airline became so successful.[39] As company historian Robert Bickers has observed, Cathay Pacific quickly adopted a strategy wherein the city benefited from its entrepot status. Intentional marketing campaigns encouraged transit passengers to empty their deep pockets, "stopping off in [Hong Kong] on the long haul to Europe – often at no extra cost to the passengers . . . [where they] shopped and

---

34 *Newcastle Morning Herald and Miners' Advocate* (NSW), April 28, 1951, 3.
35 Singapore, *Colony of Singapore Annual Report, 1955* (Singapore: GPO, 1956), 199.
36 Singapore, *Colony of Singapore Annual Report, 1959* (Singapore: GPO, 1960).
37 Koh, ed., "Paya Lebar," *Singapore: The Encyclopedia*, 406.
38 The Chinese government also took steps to create a "National Air Porty-of-Entry" with construction of an airfield at Shanghai in 1933. See Associated Press, "Modern Airport is Planned in Shanghai," *Philippines Tribune*, November 1933. The airport would see particularly heavy traffic in 1949 with the departure of Nationalist sympathizers as the Communists entrenched their control on the mainland. See *Cairns Post*, "Crowded Airport," April 28, 1949, 1.
39 Gavin Young, *Beyond Lion Rock: The Story of Cathay Pacific Airways* (London: Hutchinson, 1988), 226.

they dined, then shopped and dined again, pumping cash into its economy."[40] Cathay Pacific enjoyed a decided advantage over Singapore in this regard, as SIA would not incorporate until 1972.

The architects of modern, post-World War II Hong Kong, including Lawrence Kadoorie, understood what was at stake for Hong Kong's position as a prime tourist and stopover destination in East Asia and the Indo-Pacific. In an engrossing three-page missive, Kadoorie imagined that Hong Kong would serve as a natural attraction to both Chinese, European, and American tourists alike. "Hong Kong, with its groups of bathing sheds and villas on the coasts of the Island and the New Territories," he wrote:

> might well be the forerunner of a development similar to that of the coast between Marseilles and Monte Carlo, with its seaside villages and resorts. The fine roads and scenic beauty of the hill districts on the Island and on the mainland, as well as the coastal roads extending round the Island and from Kowloon to Ping Shan via Castle Peak, are comparable to the magnificent drive through the Esterel Mountains and the Lower, Middle and Upper Corniche roads between Nice and Monte Carlo.[41]

Unusually for the time, Kadoorie expressed strong interest in developing Chinese tourist potential in Hong Kong, largely along cultural lines. Dragon Boat Races, New Year's Festivals, visits by British royalty, and markets filled to bursting held great promise with Chinese tourists. Europeans would also be drawn to the colony's cultural, as well as natural attractions. Above all, he wrote, "Hong Kong's favourable position as one of the leading entrepots of world trade should be utilized to the full in making this Colony the 'Show Window of the Far East.'"

Key to all of this, Kadoorie well knew, was an aggressive aviation development strategy in Hong Kong. Soon after the war, he expressed concern that local officials had failed to smooth the way for Pan American Airways to operate out of Kai Tak. "Though Pan-American would very much like to use Hong Kong as a port of call," he wrote, "in view of the unsuitability of Kai Tak for landing large planes, they have reluctantly decided to use Canton." With the prospects for tourism following close on the heels of military demobilisation, Kadoorie clearly voiced what was at stake:

> My informant considers Hong Kong to be the ideal port of call in this part of the world and hopes government will take the necessary steps to enable large planes to land here at the earliest moment. He states other services are scheduled to run very shortly and that in spite

---

40 Robert Bickers, *China Bound: John Swire & Sons and Its World, 1816–1980* (London: Bloomsbury, 2020), 346.
41 Lawrence Kadoorie to Sir Patrick Abercrombie, Memorandum, "Possibilities of Hong Kong as a Tourist Centre," November 28, 1947, Hong Kong Heritage Project.

of the desire of the air transport companies to use Hong Kong as a base, unless the necessary facilities are provided, the Colony will be by-passed.[42]

Likewise, Kadoorie urged Brigadier D.M. McDougall, Hong Kong's Chief Civil Affairs Officer, "Every opportunity should be taken to show that it is Government's intention to make the Colony as important an airport as it is a seaport. Even in these difficult times, when things are not yet fully organised, I feel that we shall gain much more than we shall lose by encouraging visitors and re-establishing Hong Kong on the map." Local officials agreed with Kadoorie and Pan American Airways resumed hubbing services to Kai Tak.[43]

Kadoorie's Peninsula Hotel quickly emerged as Kai Tak's auxiliary check-in desk. In spaces today occupied by the merchants of high fashion and jewelry, airlines filled two floors of The Peninsula's lobby and arcade. In the hotel's west wing, arranged on two different floors, Pan American Airways, Philippine Airlines, Northwest Orient Airlines, TWA, Inc., and China National Airways Corporation set up shop. On the east wing, SAS, Garuda Indonesian Airways, Air America, Ltd., Japan Air Lines, Pakistan International Airlines, Malaya-Singapore Airlines, Lufthansa, Cathay Pacific Airways, Air India, Alitalia, and Canadian Pacific Airlines operated desks. Cathay Pacific Airways naturally enjoyed pride of place, located where today the concierge desk stands in the lobby.[44]

While Kai Tak served as the busiest hub in the pivot between East and Southeast Asia, Bangkok's Don Mueang Airport (opened for commercial aviation in 1925) overtook Singapore as the second busiest airfield in the wake of Singapore's wartime humiliation at the hands of the Japanese. By the spring of 1950, however, Singapore climbed back into second place, which, "With the movement of 20 aircraft a day . . . handles almost as much traffic as [Kai Tak] in Hongkong, which is the Far East's major airport." Operations in Singapore ran around the clock with "many Constellations [leaving] in the morning to reach Kolkata in the afternoon and Karachi the same evening."[45] Rebuilding from its own catastrophic wartime comeuppance, Tokyo, by 1953, challenged Hong Kong for "Gateway to Asia."[46] Farther afield, Karachi inaugurated a handsome airport in 1950 on the Orient to Europe route, complete with sumptuous overnight accommodations converted from

42 Lawrence Kadoorie to The Honourable Colonial Secretary, Hong Kong, "New Airport Development for Hong Kong," August 21, 1946, Hong Kong Heritage Project.

43 Lawrence Kadoorie to R1/11 to D.M. McDougall, Chief Civil Affairs Officer, Hong Kong, "Landing of American Aircraft in Hong Kong," October 11, 1945, R1/11, Hong Kong Heritage Project.

44 The Peninsula Hotel, "Peninsula Hotel Tenancy List," n.d., Hong Kong Heritage Project.

45 *Lithgow Mercury* (NSW), May 23, 1950, 2.

46 A.A.P, "Far East's Busiest Airport," *Launceston (Tasmania) Exa*miner, January 17, 1953, 2. On Air India hubbing at Mumbai, see *Melbourne Argus*, "Greetings to Air-India," September 11, 1956, 13.

wartime quarters, while Air India designated Mumbai as the transit point between its Asian and European operations.[47] Anticipating a national commercial aviation boom, the newly independent Indian nation made plans for thirty new airports to transform "[the nation] into Asia's most modern and best equipped ports of call for the international services from Europe and America."[48]

Indian entrepreneur J.D.R. Tata saw the writing on the wall for his fledgling new nation in the lucrative Asia to Europe market. In his autobiography he elaborated at length on the economic advantages available on a route between London and South Asian destinations: "It was clear that if . . . India were at all to enter the field of long-range international services she must do so quickly, as once foreign airlines were entrenched on all the world's best . . . rotes, India's entry would become a difficult and financially risky enterprise."[49] What set India's opportunities apart, Tata added, was location itself. He continued, "Apart from her growing importance as a great trade and travel centre, India had a commanding strategic position [astride] the only practical air route from Europe to the Far East and Australia. She was thus in a strong bargaining position vis-à-vis other countries which operated services to or through India or intended to do so."[50] Ideal location and independence aside, Air India had to contend not only with legacy European carriers, but American carriers Pan American and TWA, which "both start[ed] services to India in 1947."[51] Despite Tata's prescience, Air India settled for shared route agreements with Qantas and BOAC. The Indian government simultaneously announced an ambitious plan to condition thirty airports for domestic and foreign service. Rated the third largest network in the world, "The airports in [Mumbai], [Kolkata], Delhi and [Chennai] will be expanded and transformed into Asia's most modern and best equipped ports of call for the international services from Europe and America."[52]

Finally, at the western edge of the Indian Ocean Basin, Sharjah—located some twenty miles from present-day Dubai—appeared on the aviation radar in 1933 as a refueling station for Imperial Airways flights from Australia to Europe. Two decades later, Dubai's efforts to join the fray represented one of the most disruptive developments in the history of late twentieth and twenty-first century aviation history.

---

47 P.L.J.W., "Karachi: Largest Airport in Asia," *Mackay Queensland Daily Mercury*, January 9, 1950, 7.

48 *Wagga Wagga Daily Advertiser*, "Large Airport Network for India," November 10, 1947, 4.

49 R.M. Lala, *Beyond the Last Blue Mountain: The Life of J.D.R. Tata* (Penguin, 2017), 112.

50 Ibid.

51 Ibid.

52 *Wagga Wagga Daily Advertiser*, "Large Airport Network for India," November 10, 1947, 4.

From Military Outpost to "An Entrepot for Tourists": 1955–1980

Back in Southeast Asia, Cold War uncertainty underscored the fact that Singapore's role as an entrepot could not be taken for granted, nor was it guaranteed. While the Korean War, conflict in Indochina, and communist agitation threatened Singapore's stability, it was tensions set in motion by British mandates in the Middle East, which were also fly over zones for Qantas on its Kangaroo Route, that prompted the Australian airline, as well as U.S. airlines operating in Southeast Asia, to petition the United States government for permission to route travel between Sydney and London through North America rather than its Middle Eastern route. Indeed, conditions did not look promising in 1957, as Qantas officials noted:

> At the moment, Djakarta [, Indonesia] has denied us refueling, Singapore has just come through difficult rioting and it looks as though we will have more to come; India is demanding a new bilateral agreement with further restrictions on our fifth freedom rights; practically the whole of the Middle East area is closed to us, and we have lost quite valuable Cargo traffic . . . It appears to me that for both BOAC and QEA it is a matter of survival and we must get services operating via the US without delay.[53]

The United States' Federal Aviation Administration granted the detour in 1957, but it was rarely used.[54] Instead, unrest only prompted minor deviations in flight paths over the traditional route as well as cessation of service to hot spots in 1967 as tensions flared between Israel and the Arab world. The threat of hijacking on the route, part and parcel of concerns about Middle East instability, materialised in the 1970s, but to little effect on the Kangaroo Route.

Nevertheless, as Singapore gained footing as an independent state, local officials reimagined its role of a nexus as a strategic *commercial* advantage, particularly in the fields of air and sea trade. This also extended to traffic in transit passengers. The Minister of Communications and Works, noting the need for expansion at Paya Lebar as early as 1958, wrote, "I should like now to make a point about the value of the Government's investment in the International Airport [at Paya Lebar]. We can describe ourselves as being in the entrepôt business for passengers in much the same way that we are for goods. We receive passengers in bulk and distribute them in small parcels throughout S.E. Asia." The Minister noted that while only 15% of the passengers made quick connections through Paya Lebar, some 85% stayed on in Singapore for longer periods of time. By 1958, the number of passengers moving

---

53 John Gunn, *High Corridors: Qantas, 1954–1970* (St. Lucia, Queensland: Queensland University Press, 1988), 73.
54 Ibid., 102.

through Paya Lebar in the first six months of the year (134,124) matched the figure for all of 1954 (134,000) and contributed handsomely to economic development as "a very valuable item in Singapore's balance of payments." Officials also began to take note that associated fees collected at the airport, including "concessions and rents" (i.e., from restaurants and shops, something that would become a hallmark of Singapore airports), "[helped] to meet the costs of running the airport."[55]

While adequate at the time, Paya Lebar's construction pitted the shifting British colonial priorities against the long-term interests of Singapore. Britain only pledged enough to meet the needs of commercial aviation at Paya Lebar while local officials eyed a promising regional role for Singapore as a source of tourist dollars intimately linked to a competitive aviation hub. Furthermore, the physical relocation of PAP offices to the area around Kallang Aerodrome symbolised the rising regional and independent priorities that Singaporeans ascribed to commercial aviation and its role as a conduit for tourism development. Certainly, by the time of its independence, the early signs of aviation as a boon to tourism made themselves manifest in unfolding plans to complement manufacturing with tourism.

At the same time as Paya Lebar experienced its initial growth in the late 1950s, the government of Singapore deliberately considered the benefits of tourism development. Minister of Commerce Jumbahoy Mohamed Jumbahoy convened a Tourism Advisory Council which included Public Relations Officer, George Thomson. Thomson wanted to broaden the reach of potential customers, primarily those in the United States and Europe. The best institutional avenue for such efforts would be the PATA, a trade organisation promoting tourism development throughout the Pacific and Indian Ocean regions.

Thomson surmised that Singapore was not only a suitable stopover in Southeast Asia but provided an example of decolonisation in action. He noted:

> The Singapore Government is interested [in participating in PATA] for two reasons. Singapore, like Hong Kong, as an entrepot port cannot afford to ignore the revenue which comes from the tourist, and secondly only by receiving visitors from the other countries of the Pacific, and by ensuring that they get a balanced view at first hand can we ensure the reputation of Singapore not only as the trading centre for South-East Asia but as a community rapidly growing to full political stature.[56]

---

55 "Statement by the Minister for Communications and Works on Expansion at Singapore's International Airport for the Sitting of the Assembly, 13.8.58," Source: National Archives (Great Britain), Fonds: The National Archives (United Kingdom), Accession Number FCO 141/15129, "Singapore: British civil aviation in the Far East – migrated archives," NAS.

56 "'Tourism in the Pacific Area,' Text of Broadcast by Mr. George G. Thomson, Public Relations Officer, Singapore, over Radio Malaya on February 21st, on the Recent PATA Conference at Which He Represents the Singapore Government," February 21, 1956, Public Relations Office, Singapore,

Autonomy as a showcase, Thomson argued, was as attractive to tourists as "scenery and . . . food." In Thomson's mind, this transcended the value of tourism for its own sake and portended an opening for new markets, namely as "a very valuable medium of getting access to a wide American public for international material about Singapore."[57]

Thomson subsequently drew up justification for sending Singaporean representatives to the 1958 PATA conference, with an eye to hosting a future conference. First, Singaporean representatives could gather technical information about tourism development through "informal talks with experts." They would also be able to observe how a successful conference would be run. Additionally, their presence would raise the profile of Singapore amongst fellow regional delegates. Ultimately, Thomson averred, valuable "'know how'" would redound to Singapore through participation with the association.

The tourism council approached economic development of the industry on several fronts, including enhanced information kiosks in downtown Singapore and additional water taxis to accommodate "passengers arriving by sea."[58] Paya Lebar, however, remained their primary concern. During a May 1958 Board Meeting, the Committee adopted Resolution 205, which read:

> That after giving due consideration to the important of passenger air-traffic in conjunction with the promotion of the tourist trade in Singapore, this Committee is convinced that the tourist industry can only be properly developed by providing adequate air-transport facilities.

> That the Committee feels that nothing can ruin our reputation more than inadequate facilities at the Airport and that therefore Paya Lebar Airport as the gateway of Singapore for air-passenger traffic must present a good impression at the outset.

> That while it is fully aware of the financial difficulties the Government has to face, this Committee feels that the provision for the first-class facilities is a necessary priority expenditure as a sound investment towards the promotion of the tourist industry and that therefore

January 1956, "1. Fifth Pacific Area Travel Conference, Tokyo, 13–17 February 1956; 2. Special Joint Meeting PATA and AFETC (of IUOTO) Members – February 12," Reference Record Number: PRO 71/56, Microfilm PRO 27, NAS.

57 George F. Thomson to M. Evans, esp. February 21, 1956, "1. Fifth Pacific Area Travel Conference, Tokyo, 13–17 February 1956; 2. Special Joint Meeting PATA and AFETC (of IUOTO) Members – February 12," Reference Record Number: PRO 71/56, Microfilm PRO 27, NAS.

58 Sd. G. van der Sande to The Chairman of the Advisory Committee on Tourism, H.M. Bullard, Ministry of Commerce and Industry, Fullerton Building, Singapore, October 25, 1957, Creating Agency: Department of Information Services, "Advisory Committee on Tourism, Index of Resolutions," Microfilm DIS 15, NAS.

strongly recommends that the Government should give priority consideration to the obvious need for providing adequate passenger handling facilities at the Paya Lebar Airport.[59]

Tellingly, this resolution coincided with the hiring that very month of a tourism director by the local government. However, it made clear that local officials prioritised airport development. The proper care and handling of passengers was paramount among its duties.

Although not a republic until 1965, the Advisory Committee's recommendations illustrated a more Singapore-centric strategy for privileging airport development. It also showcased its autonomy as an attractive feature of an evolving tourism strategy. Thus, as Rachel Bok argues in a more contemporary treatment of the yet-to-be constructed Changi Airfield, proto-national priorities figured in local plans for tourism strategising.[60]

Towards that end, in 1958 the Advisory Committee on Tourism selected New Zealander R. A. Howlett, who had enjoyed success in building Fijian tourism development, as its man to raise Singapore's profile across the world. Appointed in May, Howlett got right to work. To establish a baseline, he conducted a study on the current state of tourism facilities in Singapore. Part of the review included an inventory of available hotel rooms, particularly those suitable for "Western travelers." He found that only twenty-three of the existing eighty-five hotels met with "world standards" including "being air-conditioned with own shower bath, toilet, and telephone." While hotel development comprised the chief strategy of PATA as well as that of most tourism consultants throughout the world, Howlett was mindful of the strategic advantage of Singapore's recently constructed airport, noting, "Singapore's new airport at Paya Lebar is one of the few in the Pacific now able to take all types of jet aircraft as well as normal piston engine planes. The navigational and flight side are right up-to date and passenger handling facilities are satisfactory."[61]

Howlett then launched an ambitious marketing campaign in the United States and Australia. In the United States he targeted three general magazines and three travel magazines with ads running between September 1958 and February 1959. As a consequence of efforts stateside, he noted that "results were satisfactory and many inquiries were made at the two box addresses given." Results from Australia

---

59 "Extract from Minutes of 21st Meeting of Adv. Cottee. On Tourism – 5.8.58", Creating Agency: Department of Information Services, "Advisory Committee on Tourism, Index of Resolutions," Microfilm DIS 15, NAS.
60 Bok, 2736–2737.
61 R.A. Howlett, Director of Tourism, "Pacific Review: Singapore," Creating Agency: Department of Information Services, Reference Record Number: MC&I 401–57, Folder "Advisory Committee on Tourism," Microfilm DIS 16, NAS.

were still pending. Additional visuals prepared under Howlett's direction included four-color pamphlets and six full sized posters scheduled for distribution by the spring of 1959. Finally, Howlett orchestrated the distribution of a monthly newsletter, "Signposts to Singapore," to six thousand travel agents, airlines, and travel editors. Working in conjunction with Minister of Commerce Jumabhoy, there was reason to believe that "the future of tourism [was] bright in Singapore."[62]

Concurrent with Howlett's hiring, Jumabhoy reaped the fruits that Thomson had sown two years earlier by successfully proposing and carrying off Singapore's hosting of PATA's annual conference in 1959. Comprised of representatives from throughout Asia and Pacific Island nations, the organisation targeted American tourists as potential visitors to the region. Approximately 200 delegates from 23 countries attended the five-day conference, which included seminars and presentations, as well as nightly excursions to many of Singapore's renowned attractions, including its culinary haunts, where "they feasted on satay Hokkien Mea, steaks cooked the way you like it, cold meat, barbecued chicken, spring rolls, fried rice Malayan style, ad infinitum." By day, speakers noted "the need to gear their tourist advertising towards the American visitor." Thus, "the preparations needed to keep in step in the jet age." While many nations' delegations commented on their extensive programs of hotel construction, only a few noted improvements to their airports. Ultimately, however, the pressing need for the immediate future involved adaptation to the new jet age. Executive Director George Turner "touched on the problem that the jet would cause when they arrive. Jets, he said, will cause a revolution in travel habits in the Pacific."[63]

Dependence on American tourists seemed axiomatic. Horace Sutton, Travel Editor for the *Saturday Review,* held the prospect of retreating American tourists over the heads of Asian tourism experts. He lambasted bureaucratic red-tape that inconvenienced American tourists at some-times make-shift Asian airports. If such restrictions persisted, he opined, discriminating western tourists "will find somewhere else to go where restrictions are less." "'I hope that you are stern with your Governments, strict with your hotels, relentless with your airports,'" he warned:

> The years and years of fanciful dreams built up on the minds of the Americans by Maugham and Michener, Gaugin and Loti are about to become a reality. Dun the tourist, tax him, . . . and he may very well get the idea, soon in the game that there really is no place like home. The very fact that several of us have come so far to be with you is I think a testimonial not

---

62 Ibid.
63 "The 'best ever' Annual Pacific Area Travel Association held in Singapore is over . . .," (n.d.), Creating Agency: Ministry of Commerce and Industry, Record Reference Number: MC&I T11-1-1, Folder "Director of Tourism," Microfilm MCI 0003, NAS.

only to your vitality, but to the intense interest of our readers. Many of our countrymen are waiting behind us.

These passively aggressive threats would have their effect for a time, but world wise officials including Singapore's Goh would also envision a growing number of Asian travelers.[64]

Later, the United States Department of Commerce contracted leading tourism consultant, Harry G. Clement, to make his own survey of the Asia Pacific region and to distill PATA's recommendations into an actionable strategy.[65] That the United States sponsored such a scheme, which advised its Asian participants to support hotel-construction-heavy programmes in advance of Western tourists visiting the region, should not be underestimated (for a more detailed analysis of Clement's recommendations, see Chapter 3). Once again, there was an assumption that Americans and Europeans constituted the main streams of visitors to Asia and that the most profitable approach to capitalising on such opportunities lay in the construction of large, modern hotels. While Singapore didn't ignore such advice, it marched to its own strategy of state-led airport development which would deliver higher economic gains.

By mid-1959, Goh Keng Swee, who had been charged with bringing the city-state's finances into line, caught wind of the highly publicised $300,000 public expenditure (shared by the Federation of Malaya and Singapore) on tourism development under Jumabhoy and Howlett's administration. Viewed under the lens of comparative expenses by competing tourist destinations, Howlett's efforts fell in line with the status quo. But under the pressures of balancing a budget for 1959 that was already millions of dollars in arrears, the tourism director's position made less sense.[66]

The Federation of Malaya's Minister for Commerce & Industry, Tan Siew Sin, called for a full account of Goh's intentions. At the same time, he called into question what he saw as a rash and counter-productive decision. "For my part," he concluded, "I am convinced that there are enormous potentialities in attracting to both our countries more of the lucrative travel trade in the Pacific area and that by proper promotion work tourism could be a major foreign exchange earning industry." Tan went on to question the go-it-alone approach of Goh, noting, "I am

---

64 Ibid.

65 Harry G. Clement, *The Future of Tourism in the Pacific and Far East* (Washington D.C.: Department of Commerce, 1961).

66 Goh's daughter-in-law, author of *Goh Keng Swee: A Portrait* (Singapore: Editions Didier Millet, 2015), places the budgetary crisis in a broader context, noting: "Self-governing Singapore's first-ever budget was delivered by GKS in November 1959. . . . GKS tried to ensure that the budget would be balanced, being a firm opponent of deficit spending. Education and health services took priority; total expenditure was estimated at $269 million against a revenue of $256 million." (87).

equally convinced that promotion work in this matter is best done on a pan-Malayan basis. Perhaps you disagree with this view, in which case, I for my part shall continue to press for a planned expansion of tourism promotion activities here in the Federation."[67]

Goh apologised for a decision he thought had already been conveyed by the Permanent Secretary. However, Goh took little stock in the carefully drawn up reports of the local tourism director. He noted that tourism would be transferred for the time being to the Department of Information Services "on a more modest and realistic scale."[68] Goh also closed the door on Singapore's status as a member of PATA, informing the organisation of his government's intention to withdraw not six months after Howett and Jumabhoy had hosted the annual conference. Executive Director, George M. Turner expressed "surprise" at the closure of the Department of Tourism.[69] Although he acknowledged the gravity of budgetary concerns, he expressed his confidence that "properly developed your tourist industry could be one of the greatest earners of foreign exchange for your new state." He held out hope that Goh and his PAP counterparts would reconsider their action, wishing the young nation the best. Indeed, within two years, Singapore returned to the PATA fold, though with greater oversight of its tourism expenditures.[70]

Further examination of Goh's posture towards tourism suggests that his objections had less to do with Howlett than his fundamental approach to spending money on Singapore. Ironically, subsequent (hagiographic) assessments of Singapore's tourism development have regaled Goh as one of the visionaries responsible for seeing "the potential in things as wide and disparate as animals, golf,

67 Tan Siew Sin, Minister for Commerce & Industry, June 18, 1959 to Goh Keng Swee, Tan Siew Sin to Goh Keng Swee, Creating Agency: Ministry of Commerce and Industry, Record Reference Number: MC&I 01107–59, Folder "Future of Department of Tourism," Microfilm: MCI 005, NAS.

68 Goh Keng Swee to Tan Siew Sin, June 19, 1959, Creating Agency: Ministry of Commerce and Industry, Record Reference Number: MC&I 01107–59, Folder "Future of Department of Tourism," Microfilm: MCI 005, NAS.

69 Howlett concluded his activities on August 31, 1959, finishing a term of approximately fifteen months as Singapore's tourism director. See R.A. Howlett, "D. of Tourism TO P.S. (C&I) – 4/8/59," "Leave & Departure – R.A. Howlett, Director of Tourism", Creating Agency: Ministry of Commerce and Industry, Record Reference Number: MC&I 01107–59, Folder "Future of Department of Tourism," Microfilm: MCI 005, NAS.

70 Sd. George M. Turner, Executive Director, Pacific Area Travel Association, to Goh Keng Swee, Minister of Finance, July 27, 1959, Creating Agency: Ministry of Commerce and Industry, Record Reference Number: MC&I 01107–59, Folder "Future of Department of Tourism," Microfilm: MCI 005, NAS.

tourism, music, economics, defence and education."[71] Indeed, Goh promoted creation of the Jurong Aviary Park and development of Sentosa Island as a sea, sun, and sand tourist destination in the 1960s. While Goh, indeed, saw the value of tourism, he insisted that money be spent on attractions themselves (such as Jurong and Sentosa) rather than on promotional campaigns. In the same speech where he attempted to reconcile this apparent contradiction, he also signaled where he saw the greatest tourism growth in the future:

> If I may make a suggestion, I think that there is much to be gained from studying the potential of the Asian market as against the more distant American and European markets. Americans and Europeans will come to the Orient in ever-increasing numbers and we should naturally want to get the maximum number here. But we should not forget that there are wealthy Asians in the major capital cities in the region – in Bangkok, Manila, Jakarta, Hong Kong and elsewhere. We know, for instance, that many Filipinos habitually visit Hong Kong and do their shopping there. There is no reason why they should not do their shopping in Singapore and, in addition, enjoy the experience of visiting a new place. And then, of course, there is the vast potential market in Japan. There are 100 million Japanese, and by 1975, if their economic growth continues at its present rate, their standard of living would have overtaken that of most European countries. The close attention that the Board is paying to this market is undoubtedly a right one.[72]

Thus, Goh felt that targeted, modest promotional activities should also include Asia, and that funds prioritise attractions that benefited Singaporeans as well as tourists. The recognition of increasing numbers of Asian travelers also influenced the nature of networks established by the future airline of the nation-state.

---

71 E. Shailaja Naira, *Goh Keng Swee: the master sculptor* (Singapore: SNP Editions, c2008), 14. Naira elaborates at length on Goh's apparent change of heart: "As far back as 1968, Goh was among the first to see the part tourism would play in the economy. On a 20-day trip to Hawaii, Goh studied tourism on the Pacific archipelago. When he returned, he got the Cornell University School of Hotel Management to train people in Singapore. In 1968, when tourism was put under the Ministry of Finance, Goh began a major hotel building programme. He also recommended that more money be spent on promoting Singapore abroad. To make it worthwhile for tourists to visit Singapore, he urged the quick completion of the aviary (the Jurong Bird Park) and an 18-hole golf course in Jurong. Soon, an aquarium and a zoo became part of his grand scheme of tourism. Goh got the idea for the bird park after visiting an aviary in Rio de Janeiro in 1967. He felt it would offer recreation in an industrialised city and also be a tourist attraction. And, or course, as he said, birds are cheaper to maintain than animals. . . . His ability to see the potential of tourism is evident: he enthusiastically discussed cooperation in promoting Bali as a destination, and succeeded in interesting both the World Bank and an American hotel chain in this project." (21–23).

72 Goh Keng Swee, "Speech by Dr. Goh Keng Swee, Minister for Finance at the Annual Dinner of the Singapore Hotels and Restaurants Association at the Golden Lotus, Hotel Malaysia, on Friday, 24th October 1969 at 8:00pm," https://www.nas.gov.sg/archivesonline/data/pdfdoc/PressR19691024d.pdf, accessed June 16, 2021.

Subsequently, an aviation-led tourism strategy crystallised in the minds of city planners. At the same time Goh severed ties with R.H. Howlett and the pan-Malayan approach to federation tourism development, the United Nations Development Programme (cited hereafter as "UNDP") dispatched a team of researchers, including the Dutch economist, Albert Winsemius, to assess the prospects for growth on the island. These researchers were sensitive to Goh's aversion to conventional third world developmental schemes that championed industrial substitution schemes over export-led development. Goh invited Winsemius to stay on as an economic advisor after the UNDP issued its reports on Singapore's developmental promise.

Winsemius was a prime mover behind Singapore's efforts to situate itself as a global financial center, filling a void between the closing of markets in San Francisco and their opening the next morning in Zurich. But he was also instrumental in refining Goh's unconventional approach to travel and tourism development that capitalised on transit passengers. Winsemius's approach bears repeating at length from his oral history with Tan Kay Chee:

> I developed the theory that . . . *we could develop* Singapore as a sort of entrepot for tourism. We don't have enough interesting things for tourism – to have them a week or two weeks. But what we [have], is a runway, in our airport, which will receive any airplane, and see to it that we always keep a runway which can receive the largest plane which need the longest runway. Then we can have a number of first-class hotel, elaborate to simple, but everyone of them in their group a quality hotel.

> If we can present that package, we will almost automatically have the following set up: The large planes will come from Australia, San Francisco, Europe, New York and will land at our airport. People will be tired. We'll put them [in a] bed, tuck then in a nice airconditioned room, some expensive, others relatively cheap. They can drink from the tap. They can eat what they want. The first day they will sleep and hang around . . . The second day, they will spend a considerable part of their money. A tourist is a man who has paid his ticket and has money. And that money is burning at his fingers. He will go shopping. Se we'll see to it that in the neighborhood of the not yet existing hotels, we get shopping centres.

> After a day's shopping, having spent probably half of their money already, they will go with a smaller plane or with a bus to [Kuala Lumpur], to [Sri Lanka], to the Toba Lake, sometimes to Bali. At that time you could still go to Phnom Penh, etc.

> After a week, we'll see them back in Singapore, worn-out; their stomach upset. They have eaten food they are not used to or which was not good. They have been drinking from the tap. We'll tuck them in nicely in an air-conditioned room. After two, three days, they are all set up. Then they are fit to go the other way. If they'd been to Sri Lanka, they go to the Toba Lake and if they'd been to the Toba Lake, they go to Sri Lanka.

And then once more they will come back – hot, worn-out, with a stomach upset and we will once more tuck them in nicely. Give them sleep. Get them good food that they can stand. The next day we'll put them in the long-distance airplane. Then we are [the] entrepot of tourists.

As long as we have that set-up – hotels, good water, safe food . . . shopping, plus a runway long enough to get any plane, which needs [a] long runway. We will be the entrepot for tourists. Like the entrepot for a lot of other products.[73]

Winsemius also anticipated the idiosyncratic nature of transit tourist "stays" in Singapore, which would most likely be broken up at the beginning and the end of their travel throughout Southeast Asia. While hotel development would play a central role in Winsemius and Goh's aviation-led tourism strategy, it foregrounded travel infrastructure over hotel development, not to mention the lucrative windfall from catering to the modern proclivities of transit passengers. Thus, while PATA made an all-out effort to promote private hotel development throughout Southeast Asia, often incurring inefficiencies and a glut of low-quality accommodation, Singapore adopted a focus that tailored its advantages to the transit passenger.[74]

In addition to adopting the "entrepot for tourists" approach, momentum built for an airline that would bring foreign tourists and transit passengers to Singapore. Goh's dismissal of a shared Malayan-Singaporean approach to tourism development in 1959 anticipated by a dozen years the split in Malaya-Singapore Airlines (cited hereafter as "MSA") into two carriers. While Goh does not appear to have been party to the split, his justification for breaking ties with the Malayan Federation on tourism development – namely in their divergent geographic foci – foreshadowed the division of routes and planes between the two governments by 1972. Likewise, in the division of routes, Singapore retained the international portfolio of the airline as well as the handful of Boeing 707 planes which serviced long-distance routes.

Although much has been made in popular culture about the association of the Singapore Girl and her ubiquitous *kebaya*, not only with SIA, but with the identity of the city itself, it was the decision of SIA officials, many of whom had trained with British authorities during the colonial period and under the umbrella of MSA Airlines, to purchase the latest long-haul aircraft that took advantage of Singapore's advantageous geography. Routes would be built *out* from Asia, rather than catering to traditional networks tied to New York or London. Perhaps more important than the adoption of the *kebaya*, was the rising ambition of Sin-

---

73 Albert Winsemius, Oral History, Tan Kay Chee, interviewer, NAS, Accession Number 000246, "Economic Development of Singapore," August 31, 1982, reel 12 of 18, 143–144.
74 Ibid.

gaporean officials who saw their city as a global, rather than regional, center of finance and industry.

On February 6, 1972, Sinnathamby Rajaratnam, Singapore's Minister of Foreign Affairs, spoke to the local Press Club. His remarks, later reprinted as "Singapore: Global City," imbibed the language of jet travel in its representation of the city on a worldwide stage. "The Global Cities," Rajaratnam noted, unlike earlier cities, are linked intimately with each other. "Because they are more alike they reach out to one another through the tentacles of technology."[75] He then spoke directly to role that civilian aviation played in linking these sophisticated nodes:

> some 24 international airlines operate scheduled services to most parts of the world. In 1970 there were slightly over 17,000 landings at our airport – almost treble what it was in 1960. Some 521,000 visitors passed through Singapore, some for pleasure and others on business, in 1970. We can best visualize the extent to which Singapore has become a Global City by tracing on a map the daily movements of aircraft and ships . . . like other Global Cities, [we] are nearer to one another than we are to towns which are geographically nearer. A Singaporean can get to Hong Kong quicker than he can to Kuala [Lumpur]. His major trading partners are other Global Cities rather than cities near his home.[76]

Rajaratnam's comments coincided with the structural transformation of SIA into a long-haul global airline known for safety and high-end service, as well as the city's plans to expand the receiving capacity of Paya Lebar airport for the Boeing 747, the McDonnell Douglas DC-10, the Lockheed 1011, and the Concorde. The government requested twenty million dollars from the Asian Development Bank to prepare the airport to receive wide-body jets.[77] A generation later, the Singaporean government would use the same rationale in the opening of Terminal 2 of

---

75 Sinnathamby Rajaratnam, *Singapore: Global City* (Singapore: Government Printing Office, n.d.), "text of address by Mr. S. Rajaratnam, Minister for Foreign Affairs to the Singapore Press Club on February 6, 1972," 5, https://www.nas.gov.sg/archivesonline/data/pdfdoc/PressR19720206a.pdf, accessed June 16, 2021.

76 Ibid., 9.

77 "Statement of R. Yong Nyuk Lin, Minister for Communications, at the Press Conference Held at Paya Lebar Airport on Friday 4/12/70 at 1240 Hours, 'Paya Lebar International Airport is Being Geared up for the 1980s!'" https://www.nas.gov.sg/archivesonline/data/pdfdoc/PressR19701203a.pdf, accessed June 16, 2021; The Asian Development Bank awarded the city-state $20.5 million dollars for airport expansion, the second largest loan the bank had issued to that date. The Asian Development Bank's *Annual Report* (Manila, 1971) noted the following as the purpose of the loan: "The Project to be financed will involve extension of the existing runway, construction of a passenger terminal, parking aprons, an air freight terminal, a hanger for jumbo jet aircrafts and modern communications and navigation aids . . . Singapore has a strategic location on the world's air routes. As many as 21 airlines operate services through Singapore and their number is expected to rise to 28 by 1976" (43).

Changi Airport, at that time to anticipate the arrival of "the super-jumbo of the 21st century, the A3XX."[78]

J.Y. Pillay, first chairman of SIA, identified the high risk/reward decision to purchase wide body 747s even prior to the company launching its maiden flight. He observed with some pride that SIA innovated in this way, following the lead of Japan Airlines and Qantas. As he observed at length in an oral history, aircraft procurement and not simply the *kebaya*, set SIA apart:

> We made increasingly larger investments in aircraft. We were bold. We were the first to introduce [Boeing] 747 into Southeast Asia. We branched out. We considered the world to be, shall we say, our oyster. And the world meaning really Asia and the developed world – Australia, Japan, Europe and eventually the United States and North America. Certain parts of the world where we are still not prominent, or even active in some cases, are Africa and Latin America. We don't fly at all to Latin America. To Africa, we fly to Egypt (if that can be considered part of Africa), to South Africa and to Mauritius.[79]

Low labor costs and critically acclaimed service also afforded the airline a competitive edge ahead of national air carriers saddled down with union contracts. Nevertheless, it was the adoption of long-haul jets that distinguished SIA's rise to prominence.

Singapore was not the only recently decolonised nation to envision a boost in tourism because of acquiring the jumbo jets. J.D.R. Tata salivated at the prospects of a plane with three times the passenger capacity of the Boeing 707. As his biographer noted, "J.D.R. saw the jet age through the eyes not only an aviator but one who was alert to the potential for tourism and what it would mean to have the ground facilities when the age of mass tourism began."[80] Tata subsequently collaborated with top brass in the family-owned Taj Hotel Group, "planning for the expansion of the hotel arm of Tatas, which grew from one hotel to a chain of hotels, the largest in India."[81] Nevertheless, Indian aviation infrastructure, head-scratching regulatory policies, and myopic nationalism thwarted the new nation's ability to establish a leading position along the Kangaroo Route, new aviation technology notwithstanding.

---

78 See My Goh Chok Tong, "Speech by the Prime Minister, My Goh Chok Tong, at the Official Opening of the Singapore Changi Airport, Terminal 2, on Saturday, June 1, 1991, at 11:00AM," https://www.nas.gov.sg/archivesonline/data/pdfdoc/gct19910601.pdf, accessed June 16, 2021.
79 "J.Y. Pillay: The Man Behind Singapore Airlines," in *Speaking Truth to Power: Singapore's Pioneer Public Servants*, edited by Loke Hoe Yeong, *The Singapore Story by the History-makers*, volume 1 (Singapore: World Scientific), 176.
80 Lala, 168–169.
81 Ibid.

On September 3, 1973, the Singaporean government pulled out all the stops to commemorate an important day in the nation-state's history, something that might have been more *passe* elsewhere in the world: the arrival of SIA-livery Boeing 747. The 747 had entered service at Paya Lebar two years earlier, with the arrival of a Pan American Airways flight from Los Angeles, followed by jumbos from five other airlines including British Overseas Airways Corporation, Alitalia, Qantas, KLM, and Air New Zealand. The Minister for Communications, Yong Nyuk Lin, heralded not only the arrival of the wide-body jets, but also the "world-wide net-work of offices and an operational network stretching half-way round the globe" enjoyed by SIA.[82]

At the same time, SIA officials considered an even more daring move: purchasing Concordes from the British and French governments. This was significant for several reasons. First, British Airways and SIA would jointly operate the Concorde on the Kangaroo Route. Second, in the absence of regularly serviced jets that flew non-stop between Commonwealth Australia and Great Britain, many airlines eyed development of the lucrative route. Its mere stopover in Singapore would add greatly to the city's cachet.

The Concorde made its first appearance at Paya Lebar in June 1972. Lee Kwan Yew and two of his children climbed onto the plane for the first test flight, which lasted nearly two hours and reached Mach 2 at 53,000 feet. Goh Keng Swee boarded the second test flight. British officials made a number of observations in conjunction the tests. First, W.J. Watts informed J.K. Hickman a month later that government officials, including Lee Kwan Yew, exhibited a "favourable attitude towards Concorde." Whether the Singaporeans purchased the supersonic jet or not, British officials were confident that SIA would integrate the plane into their regular service. On a more general level, they noted, "The new Singapore Airlines have done quite a lot of detailed planning for the future and see a genuine possible requirement for two or perhaps three Concordes in a few years' time."[83] British civil avia-

---

82 "Mr. Yong Nyuk Lin, Minister for Communications at the Welcoming Ceremony of the First 2 SIA-BOEING 747s at Paya Lebar Airport on Monday 3rd September 1973 @ 1600 Hours, 'SIA Joins the Jumbo Jet League,'" https://www.nas.gov.sg/archivesonline/data/pdfdoc/PressR19730903.pdf, accessed June 16, 2021.

83 "Concorde Flight News, 002 World Tour June 1972," Source: National Archives (Great Britain), Fond: National Archives (United Kingdom), Folder "FCO 14/1028: Records of the Foreign and Commonwealth Office and Predecessors. Tour of the Far East and Australasia by Concorde," Accession Number: FCO 14/1028, Microfilm Number: NAB 1354, NAS; "Concorde Visits to Singapore," W.J. Watts to J.K. Hickman, July 13, 1972, Source: National Archives (Great Britain), Fond: National Archives (United Kingdom), Folder "FCO 14/1028: Records of the Foreign and Commonwealth Office and Predecessors. Tour of the Far East and Australasia by Concorde," Accession Number: FCO 14/1028, Microfilm Number: NAB 1354, NAS.

tion officials also saw the purchase of Boeing 747s as a positive sign regarding acquisition of the Concorde, particularly "should SIA see that it would profit them to operate Concorde Singapore-London and Singapore-Sydney or Melbourne."[84]

SIA deferred purchase of the Concorde but did share services between London and Singapore beginning in December 1977. While service only lasted for three years, the experiment "was a marketing triumph for Singapore's national carrier – SIA was one of four airlines in the world to have used the super plane."[85] Doubtless, the presence of the plane also boosted the prestige and profile of Singapore as a travel entrepot amongst regional rivals and Paya Lebar as an airport worthy of supersonic arrivals and departures. The enthusiasm of Singapore's leaders for the new technology further symbolised the use of aviation advances to promote national development.

Changi: The Terminal as Tourism Catalyst (1981–2019)

The strategy of aviation infrastructure-led travel and tourism development was not unique to Singapore. The World Bank funded similar endeavors in both West and East Africa, first at Dakar in Senegal and then at Nairobi in Kenya. Terminal and runway expansion in Dakar made the West African city more attractive to airlines operating between South America and Europe. The city might benefit from "the rapid growth of tourism in Senegal and also by the increasing volume of fruit and vegetable exports to European markets."[86] Nairobi, Kenya, already enjoyed a privileged status as a tourism gateway, not only for Kenya, but also for East African destinations including Tanzania and Uganda. Without airport expansion, World Bank officials noted, Nairobi would lose its primacy as a gateway city and tourism development in Kenya would subsequently languish. As the report noted, "The new facilities will provide more efficient service and faster turnaround times for aircraft, and faster and more comfortable service for travelers. The benefits of the project will directly accrue to the Kenyan economy by allow-

---

84 "Memorandum on the Aviation Market in Singapore," Source: National Archives (United Kingdom), Folder: FV 2/847: "Sales to Singapore International Airlines. Part A," Accession Number: FV2/847, Microfilm Number: NAB 2048, Pages/frames 319–32, NAS.

85 NAS, *The 2ⁿᵈ Decade: National Building in Progress, 1975–1985* (NAS, 2010), 136.

86 See Operations Evaluation Department, "Project Performance Audit Report: Senegal-Dakar International Airport Project (Loan 867-SE)," December 13, 1977 (Washington D.C.: World Bank), 2, https://documents.worldbank.org/en/publication/documents-reports/documentdetail/559911468915318293/senegal-dakar-international-airport-project, accessed June 16, 2021 (document currently not available for download, please request).

ing the unrestrained growth of tourism, maintaining and improving the costs of Kenya's own air service."[87]

The Kenyan project closely paralleled that of Singapore's need for Changi. Like Singapore, Nairobi was a hub for an airline of intercontinental significance, East African Airways (cited hereafter as "EAA"). Like SIA, most of EAA's routes were international, either on the African continent, north to Europe, or east to Asia. Founded in 1946, by 1972, EAA had forged routes to "Hong Kong, Thailand, India, West Pakistan, Southern Yemen, Greece, Italy, Switzerland, France, Germany, Denmark and the UK." At the time of the report, EAA boasted a larger fleet than SIA, with eight jets, having carried nearly 200,000 passengers internationally the previous year. EAA's Nairobi presence attracted at least twenty additional foreign carriers to Kenya with key connections to "Johannesburg, Rome, Athens, Zurich, and Addis Ababa."[88] While Singapore's numbers quickly eclipsed those at Nairobi's expanded airport, the stakes were equally high. Singapore vied with Hong Kong, Bangkok, and Kuala Lumpur for transit travelers. Similarly, Kenya might lose prospective transit passengers to Tanzania and Uganda were the improvements at Nairobi not undertaken.[89] Singapore's aviation-led tourism strategy would also provide a palpable example for tourist-hungry emirs further west in the Persian Gulf city of Dubai.

In the case of Singapore, Paya Lebar's limited capacity, especially with inadequate space for a second runway to accommodate increased volume, not to mention future generations of wide-body jets, drove the decision to consider an array of options in the early 1970s. As a report at the time noted:

> Air transportation is of crucial importance to the economy of Singapore. The volume of air traffic handled at the Singapore International Airport increased five-fold over the last decade. Considering the prospective economic growth of the country, expansion of her trade and commerce with the rest of the world, and potential growth of tourism the air traffic is expected to continue expanding rapidly. Even on modest projections, the volume of traffic is expected to be nearly doubled over every five-year period.[90]

---

87 International Bank for Reconstruction and Development, International Development Association, "Appraisal of Nairobi Airport Project, Kenya," May 16, 1972 (New York City: World Bank), https://documents1.worldbank.org/curated/en/355021468285586310/pdf/multi-page.pdf, accessed June 16, 2021.
88 Ibid.
89 Ibid., 15.
90 ADB, "Report and Recommendation of the President to the Board of Directors on a Proposed Loan to Singapore for the Singapore International Development Project," October 29, 1970, 3, in author's possession.

Initially, the Singaporean government took a similar approach to that of Kenyan authorities, albeit in a more limited fashion: they approached the Asian Development Bank in 1970 with a proposal to extend the runway to accommodate jumbo jets. Plans anticipated completion before the end of 1971, when Qantas intended to employ jumbo jets on its routes to Europe.[91]

At the same time, Singaporean officials planned to build an entirely new terminal at Paya Lebar adjacent to the existing structure. This facility would feature twenty jet aprons complete with jet bridges.[92] With the new runway and upgraded landing facilities in place by 1973, the development bank conducted an assessment for a second loan to make Paya Lebar's expanded terminal a reality. Consultants, the Singaporean government, and private airlines instead concurred that the soaring volume of passengers and jumbo jets called for a new airport entirely. Asian Development Bank authorities noted that "[the] revised traffic projections made by NADC on the basis of the actual traffic trends experienced in Singapore in the last three years and development potentials indicate that traffic would be substantially higher than the estimates made by the Appraisal Mission."[93] R. R. Mahieu, Manager of Facility Planning Support Services for Pan American Airways reported that "strong objections were . . . raised against the ineffective design of the terminal, where airlines will be operating and processing passengers for a long time to come."[94]

Singaporean officials cancelled the second loan and opted—a decade ahead of schedule—for a new airport. By 1975, the Singaporean government informed the Asian Development Bank that is would "curtail further development at Paya Lebar;" "open the first international runway at Changi at the end of 1980;" and, "open a second parallel runway at Changi airfield by the end of 1982, on a reclamation in shallow water alongside the coast."[95] While shops and haute cuisine now attract the attention of Changi visitors, the motivation behind the new airport's construction were top-notch service and intuitive functionality. Prime Minister Lee Kwan Yew was emphatic when he discussed the strategic necessities for the success

---

91 ADB, "Appraisal of Singapore International Airport Development Project in Singapore," October 29, 1970, 21, in author's possession.

92 Ibid.

93 L.A. Hayashi to President (ADB), "Singapore International Airport Project—Back to Office Report," June 26, 1973, 4, in author's possession.

94 R.R. Mahieu to Asean [sic] Development Bank, June 25, 1973, in W.D. Kluber to H. King Hedinger, "Singapore International Airport," July 6, 1973, in author's possession.

95 ADB, "Loan No. 43-SIN: Singapore International Airport Development Project and Cancellation of Undisbursed Loan Amount," June 23, 1976, 6, in author's possession.

of the new airport. He expounded at length on the dreary state of competing airports as well as his high expectations for the new structure and airline:

> Too often, Asian airlines, and worse, disasters in Asian airports, like those in Delhi, leave passengers with the impression of sloppiness in the maintenance of equipment on aircraft and in airports. Passengers are disturbed when they see poor standards of cleanliness and, worse, slovenly service in the aircraft – no smartness in the bearing, dress and behavior of aircrew and cabin crew. They wonder if these are not also the standards of the maintenance staff and ground crew. As a traveller, I have had these thoughts. I have stressed again and again on the constant need to ensure that what passengers can see of SIA aircraft and aircrew and cabin crew, and Singapore Airport, reassures them of our attitude of mind, our zeal for tip-top standards of safety, efficiency, which prevails over all sectors . . . Maintaining these high standards is not a favour SIA workers are doing to management or the government. It is what you have to do to keep your jobs in a very competitive industry. In the airline and airport business, to stay still is to stagnate, and to stagnate is to be overtaken. We must press forward to ever higher performance.[96]

Singapore's national reports, running from 1982 through 1989, carefully tallied the growing stature of Changi, in terms of connectivity, consumer appeal, and upgrades. By 1985 it had attracted the attention of several "best airport" polls, rising by 1989 to the coveted status of "best airport in the world" according to readers of British-based *Business Traveller* magazine.

Changi International officially opened on December 29, 1981. The Minister of Defence, Mr. Howe Yoon Chong delivered an address at the inauguration. The geopolitical implications of Changi's opening were not lost on Howe, who noted the diplomatic as well as economic benefits of the new facility. He observed that the commercial aims of the new airport aligned with the economic aims of the Association of Southeast Asian Nations (ASEAN), which would stimulate "further expansion of trade, industry, tourism, banking, and finance." Likewise, Howe presaged the global shift towards Asia that Goh had mentioned some eleven years earlier. "The real growth is yet to come," Howe intoned: "Historically, the Mediterranean (the sea of the past), and the Atlantic (the sea of the present) took centuries to develop due to slow transportation provided by sailing ships and steamers. Today's wide-bodied jets promise much faster development of the Pacific Ocean (the so called 'sea of the future')." Chong envisioned the economic ascendence of the Pacific Rim, with a "loop" linking three continents. "For the present," he observed, "apart from the Japanese:"

---

96 Lee Kuan Yew, "'Address by the Prime Minister, Mr. Lee Kuan Yew, at the SIA's 30th Anniversary Dinner,' Held at the Neptune Theatre Restaurant, on Sunday, 1st May, 1977," https://www.nas.gov.sg/archivesonline/data/pdfdoc/lky19770501.pdf, accessed June 14, 2021.

> The majority of the peoples of Asia and South America living on the shores of the Pacific have yet to attain the affluence to have sufficient disposable incomes to travel for leisure. Every government strives to raise the income and living standards of its people. As they meet with success there will be further intensification of the established trunk routes through Singapore and the rapid development of the loop route.[97]

Once the oil crisis passed, he predicted, a "travel explosion" would engulf the Pacific Rim. Given these revolutionary possibilities, he closed his speech rhetorically, "Can Singapore ever afford not to have such an airport?"[98] Ultimately, these comments underscored the *global*, rather than regional, ambitions of Singapore's technocrats and administrators.

While efficiency, aesthetics and comfort remained utmost in the minds of Changi's passengers, Singapore's leading lights touted the economic benefits of aviation led travel and tourism. As early as 1977, Lee Kwan Yew posited, "When we invest in a major international airport, the primary objective is to succeed as an air junction. . . . It is important that SIA management and workers recognise this—that the primary objective is the success of the air junction. Tourism contributed 4% of our GDP in 1976, in hotels, shopping, entertainment and so on." Nearly a decade earlier, Lee championed the entrepot strategy in his dealings with regional leaders. In an interview with a New Zealand television reporter, he stated, "Well, we'd like to see more people break journey in Singapore to stay for a few days. It's a big halting place, you know, from Asia to Australasia, from East to West, and if we get people to stay for two or three days, well they can't do as much harm and we hope that will think better of Singapore after that."[99]

The first terminal accommodated up to twelve million passengers, a total quickly eclipsed by continued growth of SIA as a long-haul airline. Terminal two opened in 1990, doubling capacity. A third terminal was added in 2008, as well as a digitally automated terminal a decade later.[100]

---

97 Howe Yoon Chong, "Speech by the Minister of Defence, Mr. Howe Yoon Chong, at the Opening of the Singapore Changi Airport on Tuesday, December 29, 1981 at 3:30pm," 4–5 https://www.nas.gov.sg/archivesonline/data/pdfdoc/hyc19811229s.pdf, accessed June 16, 2021.
98 Ibid., 6.
99 Lee Kuan Yew, "Transcript of an Interview with Singapore's Prime Minister, Mr. Kee Kuan Yew, by NZBC's Hylda Bamber, Recorded in Dunedin, New Zealand, on 13th March, 1965)," https://www.nas.gov.sg/archivesonline/data/pdfdoc/lky19650313.pdf, accessed June 15, 2021.
100 Koh, ed., "Changi Airport," *Singapore: The Encyclopedia*, 92.

Changi's Asian Rivals: Hong Kong, Mumbai, and Dubai

The transition from Paya Lebar to Changi did not take place in a bubble. Singapore faced increasing competition from aviation hubs including Hong Kong, Mumbai and Dubai on the Europe to Southeast Asia and Australasia corridor. To be sure, Hong Kong was the global Asian tourism entrepot everyone else was chasing. Uncertainty over the fate of Hong Kong generally and its commercial aviation specifically, however, increased as British colonial authorities and the local government contemplated a replacement airport for Kai Tak International. As John D. Wong argues in his recent book, *Hong Kong Takes Flight: Commercial Aviation and the Making of a Global Hub, 1930s–1998*, Kai Tak was never guaranteed to retain its status as the great aviation "hinge" linking Southeast Asia, East Asia, and the People's Republic of China.[101] In 1953, for example, British crown officials debated whether it would remain a focal point of aviation in the broader region or languish as a regional destination. A memo raised a startling alternative to and verdict on Kai Tak: "It is practically certain that main trunk operators of long-haul aircraft will find restrictions at Kai Tak commercially unacceptable with modern types of aircraft. This means that such operations would have to be transferred to Manila or elsewhere, thus taking Hong Kong off the trunk routes."[102] The subsequent rise of Asian travel settled the fate of city as a global aviation hub, however. By the late 1970s, British and Hong Kong officials acknowledged that Kai Tak had reached capacity and began considering new sites for a larger airfield. Not only was the volume of flight movements increasing; wide-body planes replaced the single-aisle jets, prompting an increase in the number of passengers as well.[103]

At the same time, British officials eyed the transition from crown colony to Chinese sovereignty and considered whether the airport might be a bargaining chip in the calculus of protecting Western business interests on the island and in the New Territories. "If [the Chinese] seem reasonably receptive to the idea of exchanges on practical ways of solving the leases question and if they seem ready to talk on leases in the New Territories extending by some system beyond 1997," a secret memo ran, "then there might well be an opportunity to play on this by broaching the airport issue as an example of a major long-term project in which

---

101 John D. Wong, *Hong Kong Takes Flight: Commercial Aviation and the Making of a Global Hub, 1930s–1998* (Cambridge, Massachusetts: Harvard University, 2022), 5.
102 R. Broadbent to B.G. Smallman, "Hong Kong. Kai Tak. Aerodrome Development," September 14, 1953, COM 169/51/02, alternatively filed as CO 937/274, "Airports – Hong Kong Kai Tak," BNA.
103 Parsons Overseas Company, 6.

both sides had an interest."[104] During the flurry of discussion in the early 1980s, this meant either siting the new airport wholly within Hong Kong (more precisely, on a land reclamation project near Lantau Island) or on a location adjacent to China, which would obligate the Chinese to participate in the planning and management of the new airport. Were they to do the latter, confidential correspondence suggested, "Kai Tak would have to be kept at least in mothballs, thus greatly reducing the advantages of the scheme," in case joint administration of the new airport did not turn out.[105] Neither were British officials unaware that China was developing its own airports in the Greater Pearl River Delta, which, in effect, might compete with Hong Kong in the future.

Uncertainty pervaded discussions among British officials and in dialogue with the Hong Kong government and People's Republic of China. It would be some time after Margaret Thatcher and Deng Xiaoping agreed to the Hong Kong SAR arrangement, which purportedly safeguarded Western economic and political practices until 2047, that the People's Republic of China and British officials agreed upon building the airport at Chek Lap Kok. This 1991 agreement ensured that the Hong Kong government would cover the costs of the new airport but authorise expenditures for future alterations, albeit with Chinese approval.[106] Thus, while an agreement had been reached regarding airport location, the future of the airport under Chinese control remained as uncertain as the future of the city-state itself. Then, as now, one of the concerns was that Hong Kong's hub status might be challenged in the future by neighboring Chinese airports.

Cathay Pacific executives expressed guarded optimism. In 1994 and 1995, Peter Sutch gave a series of talks on the future of Hong Kong. The Chairman of the Swire Group touted Hong Kong as the gateway to mainland China, as well as the rest of Asia. Kai Tak reflected this reality with its near capacity schedule of airplane movements. Anticipating completion of the new airport, he predicted that "[t]hereafter, Hong Kong will have almost unlimited growth potential as *the* aviation hub for Asia. After all Hong Kong could not be better located at the heart of Asia, four and a half flying hours from half the world's population."[107] Drawing

---

104 See R.D. Clift to Sir Murray MacLehose, Governor and Commander-in-Chief, Hong Kong, September 30, 1980, Hong Kong 182/1 file, "Possible New Airport for Hong Kong," FCO 40/1181, BNA.
105 Ibid.
106 British Foreign and Commonwealth Office, "The New Airport Hong Kong Airport," Background Brief, September 1991, FO 973/665, British National Archives. Hong Kong officials did their due diligence in briefing the Chinese government on each step in the building process, including seeking approval for financing during construction in the early 1990s. See S.H. Broadbent to Mr. Rickets, June 12, 1992, FCO 30/4500 folder, "Provisional Airport Authority," File HKA 182/1, BNA.
107 Peter Sutch, Chairman, John Swire and Sons (HK), "Hong Kong: A Springboard to China and the Asia-Pacific," Tokyo, Tuesday, September 26, 1995, "Copies of speeches given by Peter Sutch,

a tighter connection between hub and airline, he observed in another setting, "There are other airports being built or planned in the region which are trying to promote themselves as the ideal gateway to Asia, but none has the advantage of Hong Kong on China's doorstep, and at the very heart of Asia."[108] Philosophically, Sutch believed that the liberalisation of travel throughout mainland China spoke to growing freedoms, a potential sign of seismic political and economic changes ahead. "I am struck by how dramatically freedoms have grown – in particular the freedom to travel," he noted. "Our experience inside the Swire Group as China's aviation market has opened up is striking evidence of this, with our . . . subsidiary now serving 14 cities across the mainland."[109]

But the future would be more complex, even for Cathay Pacific. In the early 1980s, when serious discussions about Chek Lap Kok began, Cathay Pacific executives breathed a sigh of relief when the airport was shelved "until the long-term future of Hong Kong is known."[110] The airline was apprehensive about the cost of relocating its offices to a new location. While they faced the reality that such a position was untenable as "more and more existing [planes] will be switched from L10-11, DC-10 and Airbus to 747 to cater to increased passenger demand," a secondary concern involved the yet unknown demands of Chinese civil aviation. "It is not impossible that one day CAAC [(Civil Aviation Administration of China)] will become commercially aware of its almost unique potential . . . into and out of Hong Kong to Europe, [and] they would undoubtedly want to put a daily widebody to Shanghai and a daily widebody to [Beijing] right in the middle of the peak, and who is going to refuse them?"[111]

Furthermore, the retooling of Chinese fleets to widebody jets would not prevent the Chinese from building their own airports to compete with Hong Kong.[112] By the late 1980s, however, Cathay Pacific's future was intertwined with that of Chinese aviation. The Swire Group tied-up a joint venture with the Chinese government (CITIC) in the regional carrier Dragonair, which focused on routes be-

---

Head of Cathay Pacific Airways until 2002, and Taipan of John Swire & Sons, Ltd.," GB 102 JSS/13/19, School of Oriental and African Studies Archives (cited hereafter as SOAS).

108 Ibid., 19.

109 Peter Sutch, Chairman, "Riding the Dragon – the Swire Group Beyond 1997," 1997, Empire Club, Toronto, Canada, Tuesday, November 15, 1994, "Copies of speeches given by Peter Sutch, Head of Cathay Pacific Airways until 2002, and Taipan of John Swire & Sons, Ltd.," GB 102 JSS/13/19, SOAS.

110 D.R. Bluck to H.J.C. Browne, "New Airport," July 15, 1982, GB 102 JSS/13/2/12, SOAS.

111 R.T. Stirland, General Manager Airline Planning to D.R.Y. Bluck, Chairman, "Kai Tak," May 18, 1982, GB 102 JSS/13/2/12, SOAS.

112 D.R.Y. Bluck, "Aviation Advisory Board – New Airport," May 11, 1982, GB 102 JSS/13/2/12, SOAS.

tween Hong Kong and mainland China.[113] A shared maintenance facility capable of accommodating widebody jets had also been installed in nearby Xiamen. The hope was that commercial aviation and business culture generally would be complementary rather than adversarial. "There is no question that Hong Kong will continue to play [a] vital role in the southern part of [China] as Shanghai will continue to stimulate growth in the east," Cathay executive Simon Heale observed. "I am just delighted to work for a company that manages an airline that flies between Hong Kong and Shanghai. There is no doubt that this will become one of the world's busiest pairs of the future."[114]

In terms of Indian aviation, J.D.R. Tata aspired to be the best in the industry. Effectively ignored by British officials while Imperial Airways established a scheme linking London to Sydney in the 1930s, Tata found renewed hope at the outset of independence. This included securing the latest aircraft as well as offering superlative customer service. Tata tirelessly pursued the best aircraft and prudently so. He demurred at the suggestion his family underwrite commercial aviation on their own, hoping that the Indian government would see the virtues of a national scheme. It would be in the post-World War II era that his vision came to pass. "I had almost given up hope," he wrote to Sir Frederick Tymms, "and was getting ready finally to cancel our provision order for the three Constellations when, at the last moment, the Government of India acted with unusual quickness of decision and accepted our scheme almost in toto."[115] A long-hoped for route from India to London materialised.[116] He saw the advantages of bigger aircraft after World War II and eventually advocated for the acquisition of a fleet of Boeing 707s and then 747s.[117] However, in his later days as Chairman of Air India International, Tata urged caution when approached about purchasing the

---

113 Simon Heale, "The Corporate Strategy of the Swire Group towards China," June 7, 1995, "Copies of speeches given by Peter Sutch, Head of Cathay Pacific Airways until 2002, and Taipan of John Swire & Sons, Ltd.," GB 102 JSS/13/19, SOAS.

114 Ibid.

115 JDR Tata to Sir Frederick Tymms, December 27, 1947, in Arvind Mambro, ed., *Letters* (New Delhi: Rupa & Company, 2004), 186. Tata also facetiously praised his competitors for helping his cause: "Fortunately, the poor service given by the American airlines serving India, and the obsolete equipment used by B.O.A.C., will greatly reduce our initial competition, in addition to which we have over these many years managed to create a reputation and passenger loyalty towards us which will be of great help." (186–187).

116 See a note of congratulations on this accomplishment in Mrs. Rodabeh Sawhny to Tata, October 17, 1957, in Mambro, 203.

117 Tata lauded the wide adoption of the Boeing 747 by other airlines as proof of Air India's wise decision to build a fleet of the jumbo jets. See Tata to Peter Mingrone, August 9, 1974, in Mambro, 231.

Concorde. In hindsight, he felt good about holding off until the new model had been proven in a second generation. In the meantime, the price of jet fuel spiked due to the OPEC embargo and flying the Concorde proved problematic as sonic booms caused growing objections to the advanced technology.[118]

As far as customer service was concerned, Tata intuited the refined tastes of his passengers, particularly on long-haul flights. Among his published letters, twenty percent of those dealing with commercial aviation touch on in-flight service experienced on Air India International or its competitors. He was mindful of the unique challenges posed by long-haul travel, particularly in the wake of advances in jet technology. As his competitors outfitted their new jets with as many seats as possible, Tata insisted that Air India retain its "slumberettes," a forerunner of the lay flat first-class seat. "I am glad that you, like Gene Black, approve of Air-India's slumberettes," an emboldened Tata wrote to George Woods, future President of the World Bank:

> The fact that the jets are faster does not mean that people are likely to spend less hours in them. For instance, some travellers between the U.S. and India, who used to break journey in Europe, may now travel straight through. Even in a jet, a flight from London to Sydney or London and Tokyo will still take around thirty hours, and thirty hours upright in a chair can be as uncomfortable in a jet as in a piston engined plane[119]

He anticipated this in the 1950s and remained convinced that he had made a good decision when comparing first class offerings thirty years later during flights on Swiss, Lufthansa, and British Airways.

Finally, to a limited degree, Tata also recognised the importance of connections between the three pillars of the hubbing effect. He sometimes dropped in on airports to inspect the quality of ground services. On one occasion, he left Mumbai's Santa Cruz International Airport aghast at the dysfunctional luggage delivery system. "I found the conditions indescribably bad at the arrival end, and only slightly less so on the departure side."[120] He found "the situation . . . nothing less than inexcusable for an international airport of a city like [Mumbai]."[121] Thus, while he wrote little about airports explicitly, this connection between customer comfort and the hallmarks of an exceptional airport drew his ire. Likewise, Tata sensed the strong connections between aviation and hotels. Wearing hats as both Chairman of Air India International and Taj Hotel Group, he took issue with a letter he received from Mohan Singh Oberoi, owner of a rival hotel chain, accus-

---

118 See Tata to the Earl of Kimberley, April 20, 1976, in Mambro, ed., 233–234.
119 J.D.R. Tata to George Woods, November 20, 1959, in Mambro, ed., 145.
120 Tata to Air Marshall Y.V. Malse, January 1, 1975, in Mambro, ed., 159.
121 Ibid. 160.

ing him of funneling too few passengers to Oberoi's two Mumbai hotels. Tata went to great lengths to refute such claims, as well as to instruct his rival of the synergies achieved by pairing air travel and hotel services. He also possessed wide enough vision to recognise the benefits to local tourism brought about by exceptional hotels.[122]

While we have touched on the origins of the Dubai airport in 1959 as an outgrowth of British colonialism, it was not until 1983 that shopping became *the* priority of Dubai officials for transit passengers. That strategy came from an unexpected source. In the age before Boeing 707 jets, Shannon, Ireland, was an obligatory refueling stop for trans-Atlantic flights. Longer-range aircraft, like the Boeing 747, made the fueling function of the airport redundant.[123] Thus, Shannon reimagined its relevance for long-haul flights. In addition to securing the unique role as a pre-landing customs clearance site for passengers headed towards the United States, it upped its game as a duty-free destination boasting Waterford crystal and cable-knit sweaters that were the toast of the tarmac amongst airport retailers.[124]

In 1983, Mohi-Din Binhendi, Director General of Dubai Civil Aviation, chanced upon brochures of Ireland's duty-free concept, and having considered enhanced retail offerings at his own airport, pitched the concept to his superiors. Sheikh Mohammed bin Rashid Al-Maktoum thought bigger: "'We were not measuring ourselves against the Gulf, nor the Middle East. Even in the early 1980s our yardstick was the best airports in the world. There was a long way to go, of course, but that was the intention.'"[125] Binhindi and his team made their way to Shannon to scout out the duty-free concept, hire a group of Aer Rianta officials on a temporary basis to set up a similar (thought much larger) enterprise in Dubai, and open for business the same year, five days before Christmas.

As had happened with Dubai seizing a stranglehold on Gulf airfield primacy with its "open skies policy," the Emirati's duty-free team negotiated with Aer Rianta to delay consultation with competing airports in the region. Next came support from the Dubai government to launch an ambitious programme, which would be the centerpiece of the airport. As historian Graeme Wilson has written:

> Dubai wanted to greatly increase its current throughput of passengers, which totaled around three million a year in 1983. The government wanted to tap into the transit market

122 See Tata to Rai Bahadur M.S. Oberoi, August 19, 1974, in Mambro, ed., 392–393. Making the connection between air travel, hotels, and tourism, Tata writes, "The Taj has invariably responded to Air-India's approaches for special concessions and facilities for tourists brought to India by Air-India."

123 Graeme H. Wilson, *Fly Buy Dubai*, 33.

124 Ibid., 44.

125 Ibid., 69.

between East and West – their wish was that a duty-free operation, similar to that successfully operating in Shannon, would attract new airlines and thus a whole new cache of transiting passengers. The duty-free operation had to be not only first class but so attractive it would draw like a magnet and turn Dubai International into a 'must visit' destination.[126]

The challenge was to source products of global appeal (such as perfumes and cigars, for example) as well as those that piqued the interest of different traffic streams passing through Dubai. Gold drew the greatest interest, due in part, perhaps, to the large number of passengers originating from the Indian sub-continent. In the end, even before Dubai boasted the largest shopping malls in the world, its airport had virtually become a shopping center. "Over 15,000 products were on the shelves of Dubai Duty Free's first ever shops," Graeme Wilson wrote. "The stock ranged from music, and video tapes, electronics, toys, sporting foods, ladies' apparel, gents' clothing, leatherwork, fragrances for men and women, clocks and watches, jewelry, gold, pearls, textiles, souvenirs from Dubai, confectionary, delicatessen items, international gifts, a children's shop, tobacco items, and liquor."[127] Tourism quickly followed.[128] The shopping draw drove the rise of leisure travel and aviation to Dubai. "It has become a well-known fact of aviation history—not just regionally but internationally—that the marriage between the airline, the airport, and the duty free in Dubai has been an incredible success and the envy of other countries, many of whom have aspired to emulate the triple winning formula."[129] In this sense, Dubai was as much an innovator as Singapore and Hong Kong in drawing tourists to airport hubs.

The Jewel Project: The Terminal as Tourist Destination

Strategically, one-time SIA Chairman Lom Chin Beng noted, "the Government has said that if it has to decide between protecting Changi's air hub status and losing SIA's standing in the new aviation landscape, it will choose Changi."[130] As of 2013, when Lee Kwan Yew's son, Prime Minister Lee Hsien Loong, announced the con-

---

126 Ibid., 93.
127 Ibid., 134.
128 Wilson writes, "The first real tourism drive began in Dubai in late 1984 when the emirate had only two dedicated beachside tourist hotels. However, the city also boasted severally internationally recognised business hotels including the Sheraton, Hyatt, Hilton, and InterContinental, and they were well patronised by the hordes of businessmen [and shoppers] who were flocking into Dubai in search of new opportunities, [bargains] and contracts" (148).
129 Ibid., 173.
130 Ken Hickson, *Mr. SIA: Fly Past* (Singapore: World Scientific, 2015), 158.

cept of "Jewel," he noted, "The airport and all the things which are connected with the airport, all the related services, they provide a lot of jobs in Singapore. I would not ask you to guess how many. But it is 163,000 jobs in Singapore, 6 per cent of the GDP and it is all levels of society."[131] Ultimately, however, the benefits of Changi to passengers and Singaporeans alike endeared the facility to its many patrons. "What is Changi Airport?," Lee Hsien Loong opined in his 2013 speech, "To travellers – an icon of Singapore. To Singaporeans – a welcome landmark telling us that we have arrived home. To me it is a part of the Singapore identity – a symbol of renewal and change."[132]

The lines between infrastructure and superstructure—amenities appealing to the care and comfort of tourists—began to blur at Changi, reaching its culmination with the Jewel facility in 2019. Within a year of Changi's opening (in 1981), state reports ran, "Changi Airport has enhanced Singapore's image as a shopper's paradise. Sales at airport shops were in certain cases as much as three times more than those at Paya Lebar Airport."[133] A medical clinic and "in house colour television system" were installed the following year.[134] An airport "meat shop" joined the nine restaurants and thirty stores at Changi in 1984.[135] In 1985 conceptual additions to the airport elevated Changi to a level unmatched elsewhere in Southeast Asia. At that time, a children's "play corner" was opened in the transit area.[136] In 1986, the first theatre was installed in Terminal 1. For passengers with more refined taste, during that same year, "two new cocktail lounges with pianist performance were introduced in the departure/transit lounge."[137] 1987 witnessed the inauguration of a highly anticipated business center, car valet service, as well as publication of an airport magazine (airport-related melodramas would be broadcast online in the 2010s for the enjoyment of sometime Changi denizens

---

131 Prime Minister Lee Hsien Loong's National Day Rally 2013 (English), August 18, 2013, https://www.pmo.gov.sg/Newsroom/prime-minister-lee-hsien-loongs-national-day-rally-2013-english, accessed June 15, 2021.

132 Ibid.

133 Ministry of Communications and Information, *Singapore 1982* (Singapore: Ministry of Communications and Information, 1982), 59.

134 Ministry of Communications and Information, *Singapore 1983* (Singapore: Ministry of Communications and Information, 1983), 130.

135 Ministry of Communications and Information, *Singapore 1984* (Singapore: Ministry of Communications and Information, 1984), 139.

136 Ministry of Communications and Information, *Singapore 1985* (Singapore: Ministry of Communications and Information, 1985), 162–163.

137 Ministry of Communications and Information, *Singapore 1986*, (Singapore: Ministry of Communications and Information, 1986), 117–118.

around the world).[138] Finally, in 1988, one year shy of its first award as the world's "best airport," Changi introduced free city tours for transit passengers, while local hotels, shops, and restaurants gamely offered transit passengers discounts at local businesses.[139]

By 1990, with the opening of the second terminal at Changi, Australians and Hongkongers took note of Changi's exceptional facilities. Terminal 2 doubled the capacity of Changi, with its "100 shops, 20 restaurants, two fitness centres, hairdressing salons, supermarkets, banks, 14 currency exchanges, post offices, and a business centre." It was not the innovations alone that raised eyebrows; it also raised expectations for potential travelers and rival hubs. Australians noted the disparity between check in facilities at Singapore and Sydney. The aviation editor for *The Sydney Morning Herald*, Tom Ballantyne, saw this as nothing more than an effort "to become the major aviation hub of Asia, eclipsing Bangkok, Hong Kong and Tokyo."[140]

Hongkongers voiced concerns about Kai Tak losing Asian hub preeminence. Singapore offered an enviable alternative. In a *South China Morning Post* screed, entitled "Air of Disgrace over Kai Tak," Kevin Sinclair raised collective angst to an existential plane:

> When you die and go to heaven, Saint Peter will throw open wide the pearly gates and usher you inside. Once in paradise, you will look around in appreciation, whistle and say: 'Oh boy, this is terrific. It's almost as good as Terminal Two at Changi Airport.' . . . Singapore's splendid new facility is the most magnificent on earth. Not a cent has been spared in making this marble, glass and chrome edifice a palace to aviation as the island republic seeks to clinch its hold as air hub of Southeast Asia.

A comparison was not long in coming. Quite simply, Sinclair lamented, "Kai Tak is a disgrace and Hongkong should be thoroughly ashamed of it."[141] As Hong Kong officials dithered, Lufthansa also made its displeasure known, threatening to move operations to neighboring Macau if further delays postponed the construction of proposed Chek Lap Kok airport. What was at stake was not Hong Kong's viability as a serviceable airport, but instead its strategic, geopolitical importance, which

---

138 Ministry of Communications and Information, *Singapore 1987* (Singapore: Ministry of Communications and Information, 1987), 126.

139 Ministry of Communications and Information, *Singapore 1988* (Singapore: Ministry of Communications and Information, 1988), 128–129.

140 Tom Ballantyne, "Singapore Builds the World's First City-in-an-Airport," *The Sydney Morning Herald*, November 22, 1990, Factiva.

141 Kevin Sinclair, "Air of Disgrace Over Kai Tak," *South China Morning Post*, June 17, 1991, Factiva. Also see Bernard Fong, "Travellers Unhappy with Kai Tak Services," *South China Morning Post*, November 26, 1990, Factiva.

might be lost to the former Portuguese colony "'as a real Asian hub for European airlines and also as an intermediate stop airport.'"[142]

True, Singapore's new amenities were not "Asian," per se. They were modern and configured in such a way that they became synonymous with Singapore's material identity. Innovation in aviation infrastructure, or the latest creature comforts at Changi, became one and the same with the experience passengers anticipated when flying through this "entrepot for tourists."

Similarly, passengers came to associate Singapore with the latest aviation technology and ever-expanding routes. Fleet modernisation and connectivity on the SIA side kept pace with the roll-out of new amenities at Changi. All South Pacific routes were covered by B-747's by 1982, with new service to Abu Dhabi introduced the same year. In 1983 SIA ordered eight B747s "with stretched upper deck[s]," while mothballing 26 aircraft, proof of its commitment to maintain a young fleet. Additional "Big Top" B747-300s were ordered in 1983–84 and were integrated into service on the Australian and London routes, which previously enjoyed some connectivity via the Concorde. 1985 saw the addition of Beijing as a destination, as well as openings in Vienna and the African island of Mauritius.[143] SIA continued to commit its resources, 88% in 1987, to new equipment, largely in new aircraft (B747-300 "Big Tops" and Airbus A310s). Global expansion continued apace with new flights to Frankfurt and Zurich in Europe and San Francisco in the United States.[144]

New terminals with amenities including theatres, verdant walls and florid floor landscaping, not to mention a butterfly garden, were not enough in the minds of Changi and Singaporean officials to retain global accolades. Even as the airport awaited construction of Terminal 3 in 2001, Yeo Cheow Tong repeated the mantra of continuous improvement at an airport reception. "Remaining number one is not easy," he reminded, "As you are already well aware, competition in the region is heating up. A number of new airports have already been opened, or are under construction, equipped with state-of-the art facilities."[145] Nearing the end of his service as Mentor Minister, Lee Kuan Yew was not above "naming" regional

---

142 Giselle Militante, "Warning on Airport Delay," *South China Morning Post*, March 16, 1991, Factiva.

143 Ministry of Communications and Information, *Singapore 1986* (Singapore: Ministry of Communications and Information, 1986), 117–118.

144 Ministry of Communications and Information, *Singapore 1987* (Singapore: Ministry of Communications and Information, 1987), 128–129.

145 Yeo Cheow Tong, "Speech by Mr. Yeo Cheow Tong at the Annual Airport Reception on January 13, 2001," https://www.mot.gov.sg/news-centre/news/Detail/Speech%20by%20Mr%20Yeo%20Cheow%20Tong%20at%20the%20Annual%20Airport%20Reception%20on%2013%20January%202001/, accessed June 15, 2021.

competitors that challenged Changi's global leadership. As if sounding the warning bell, Lee observed:

> It would be a mistake to believe that past achievements will guarantee our continued success in the years ahead. Competition amongst air hubs is increasing with new mega-airports in the region like the Beijing Capital International Airport, Shanghai's Pudong Airport, Incheon International Airport [in Seoul] and Dubai International Airport. Our competitors are catching up, with some beginning to equal, if not outperform us in certain areas.[146]

Lee's son, Prime Minister Lee Hsien Loong, affirmed his father's concerns four years later in his National Day Rally Speech of 2013, noting that "other airports in Southeast Asia are expanding to take advantage of these opportunities. [Kuala Lumpur International Airport], they are planning to service 100 million passengers per year. Bangkok Suvarnabhumi (Airport) also aiming for 100 million passengers a year and both of them are geographically better placed than Singapore to be the hub of Southeast Asia." Closer proximity to Europe gave its competitors a slight edge but could not match the continuous improvement in Singapore, because, as the prime minister exulted, "They are not Changi airport! That makes a difference."[147]

Under Lee Kwan Yew and Goh Keng Swee's watchful gaze, primary responsibility for planning at Changi fell to the airport's management team. Lines of institutional authority were clearly defined. Nevertheless, other entities, including the Singapore Tourism Board (cited hereafter as "STB") coordinated their future developments with the airport in mind. In a 1986 planning report commissioned by the Board, consultants encouraged Singaporean officials to use space in the airport to showcase art as had been done in the recently completed San Francisco International Airport. If a rationale were required, the consultants suggested, "The estimated one million transit passengers who stay for extended periods (2–7 hours) would be justification enough."[148] While conservative in nature, the proposals were offered "to make the airport experience more enjoyable."[149] Years later, the STB imagined the totality of Singapore as "The World's Biggest Transit

---

146 Lee Kuan Yew, "Speech by Mr. Lee Kuan Yew, Minister Mentor, at Launch of the New CAAS and the Changi Airport Group, July 1, 2009, 4:00pm at Changi Airport Terminal 3," https://www.pmo.gov.sg/Newsroom/speech-mr-lee-kuan-yew-minister-mentor-launch-new-caas-and-changi-airport-group-01-july, accessed June 15, 2021.
147 Lee Hsien Loong, "Prime Minister Lee Hsien Loong's National Day Rally 2013 (English), August 18, 2013," https://www.pmo.gov.sg/Newsroom/prime-minister-lee-hsien-loongs-national-day-rally-2013-english, accessed June 15, 2021.
148 Singapore Tourism Board, *Tourism Development in Singapore: a Report* (Singapore: Singapore Tourism Board, 1986), v–44.
149 Ibid. v–45.

Lounge." Touting the scores of airlines arriving from nearly two hundred cities across the world, the Board acknowledged the appeal of Changi's amenities, but also promised tourists "convenient access to the many attractions outside the airport for those with longer stopover time, essentially turning Singapore into a huge transit lounge."[150]

Since the inception of empires, state leaders have closely guarded technologies or sources of information that prove decisive in achieving superiority over rivals. The Portuguese did it with superior maps in the early modern era of exploration. The Dutch followed with superior naval technology that facilitated faster construction of ships. Much later, during World War II, the United States acted emphatically with atomic technology to close the war. In his 2013 National Day Rally Speech, Prime Minister Lee Hsein Loong turned to airport infrastructure to remain one step ahead of competing Asian mega-airports, but maintained an air of secrecy when he announced the redevelopment of an airport parking lot. "We are going to replace [it] with what we have codenamed 'Project Jewel.'" Intimating the magnitude of such infrastructure that would serve as a travel *destination*, he continued, "it will have shops, restaurants and a beautiful indoor garden. So we have Gardens by the Bay, this one is Gardens at the Airport." Following in the time-honored tradition of Goh Keng Swee's philosophy of local benefits for tourism development, Lee continued, "[the Jewel is] not just for visitors but for Singaporeans too – families on Sunday outings, students maybe studying for exams, newlyweds taking bridal photos."[151]

Airport authorities in rival hub cities, including Hong Kong, quickly took notice of Singapore's plans to position itself as *the* preferred global transit point in the region. In 1998 Hong Kong ushered in a new airport, the engineering island marvel Chek Lap Kok. By 2017, however, the number of transit passengers moving through Hong Kong dipped. In response, the Legislative Council Secretariat's Research Office published a thorough, thirty-two-page study, "Policy Measures to Enhance Airport Competitiveness in Selected Places," which compared Chek Lap Kok with Singapore's Changi and Dubai's International Airports. It was not simply that the number of transit passengers had declined, but also the airport's *reputation* with

---

150 Singapore Tourism Board, https://www.stb.gov.sg/content/stb/en/media-centre/media-releases/the-worlds-biggest-transit-lounge.html, accessed January 13, 2022. T.C. Chang has summarized the Singapore Tourism Board's development strategies, as well as later diagnostic studies, including *Tourism 21*, Compass 2020, which were not available to this author. See T.C. Chang, "Sustaining Singapore Tourism Through the Years: Policy, People, Place," Lawal Mohammed Marafa and Chung-Shing Chan, eds., *Sustainable Tourism in Asia: People and Places* (Newcastle on Tyne, UK: Cambridge Scholars Publishing, 2019), 188–199.
151 Ibid.

passengers. Since 1998, Chek Lap Kok had been voted the world's best civilian air-port, the LegCo report noted, "eight times up to 2011." However, with the ascendancy of Changi and the Persian Gulf airports, including Dubai, "[Hong Kong International Airport] has slipped consecutively to the 5th position in terms of the global ranking of 'best civilian airport' in 2017."[152]

Competition was real, the report emphasised. "Hong Kong cannot be immune to fierce competition from other airports," the study ran, "long-haul passengers may choose to transit-transfer at other global hubs like Dubai, [Singapore] and Seoul, instead of Hong Kong." While the city-state had taken some measures to staunch the flow of transit passengers, including the elimination of an air departure tax and "discounts for visiting certain attractions such as Hong Kong Disneyland," its rivals upped the ante. "Snooze cubes" were cited as one advantage Dubai offered to haggard passengers, where they could unwind with wi-fi ports in a modicum of privacy. Transit passengers also enjoyed "access to a swimming pool equipped with gym, sauna and shower facilities for a small fee."[153]

But it was Singapore's Changi where the real competition lay. City-state fathers had the foresight to anticipate growth, something that caught most of its competitors flat footed. "'If you look at Hong Kong, Bangkok, Jakarta and Manila, through a combination of infrastructure constraints, political decision-making, environmental constraints, all of them, the airport doesn't get the capacity until long after they should have it. And Changi has been exceptionally good at that,'" a Singapore-based consultant noted.[154] What's more, passenger fees had been slashed by two-thirds. Other amenities bear repeating at length, as if they were state-level intelligence:

> to make their short stay comfortable, the airport authority offers free facilities such as (i) snooze lounges; (ii) computers with internet connection; (iii) 24-hour free movie theatre, and (iv) video game zone, along with 24-hour restaurants. Connecting passengers may also visit three transit hotels inside the airport, either for a nap as short as a few hours or taking a shower at affordable charges without passing through the immigration clearance. Thirdly, a new terminal complex called Jewel acting more like a multi-purpose shopping and entertainment mall is scheduled for opening in the early 2019 [sic]. With a 40-meter indoor waterfall and a largest indoor garden in Singapore, it could provide much entertainment to [transit] passengers staying at the airport.[155]

---

152 Chun-ho Yu, Legislative Council Secretariat Research Office, "Information Note: Policy Measures to Enhance Airport Competitiveness in Selected Places," https://www.legco.gov.hk/research-publications/english/1718in01-policy-measures-to-enhance-airport-competitiveness-in-selected-places-20171027-e.pdf, accessed January 19, 2022.
153 Ibid.
154 Danny Lee, "Airports Battle it Out," *South China Morning News*, November 3, 2017, Factiva.
155 Chun-ho Yu.

The study recommended that Chek Lap Kok "provid[e] entertainment options at both the airport terminals or downtown at a discount."[156]

On the opposite side of the Indian Ocean Basin, Dubai's city-state airline, Emirates, and its rapidly expanding International Airport, did not necessarily challenge Singapore's position as an entrepot as much as did Hong Kong, but instead took inspiration from a similar strategy of a hub-cum-entertainment center appeal with transit passengers. By 2008, city officials heralded the opening of the world's largest terminal—Terminal 3—in its' international airport, followed by a mammoth concourse dedicated specifically to the Airbus A380 workhorse. Dubai gave a nod to what Singapore had accomplished in Southeast Asia. An independent report noted in 2011: "Dubai's openness to foreigners and its positioning as a hub for the Middle-East is unique in the region," the study observed, "which could enable it to serve transit tourists as Singapore has done for Malaysia, Thailand, Vietnam, and Indonesia."[157] Furthermore, as writer Ben Simpfendorfer noted, more was at stake for Dubai than hubbing for the Middle East or Europe-bound South Asians:

> Dubai offers Asian companies an attractive business hub, especially for those dealing with North and East Africa, where the populations are often Muslim, speak an Arabic dialect, and have strong historical ties to Dubai or the Middle East. For instance, in spite of all the talk about China's growing trade with 'Africa,' some 26 percent of its total trade with the continent was with North Africa in 2012.[158]

Additionally, a team of Australian-based researchers reported, "Dubai is more aggressive and consequential as seen in Emirates' decisive stand-alone strategy . . . With the inauguration of flights to Sao Paulo in October 2007, Emirates is now the only carrier in the world that serves all continents with its own aircraft."[159] The

---

156 Ibid. Three years later, the Hong Kong Airport Authority launched their own infrastructure salvo across the Asia-Pacific and Indian Ocean Basin, announcing the development of SkyCity, a "retailtainment" center, complete with over 800 shops and restaurants (more than three times the number at Changi's Jewel), a Marriott hotel, and entertainment options, set to roll out by 2025. See https://www.skycityhongkong.com/en, accessed January 20, 2022.

157 Aldi Haryopratomo, Sanja Kos, Lavin Samtani, Sheela Subramanian, and Jay Verjee, "The Dubai Tourism Cluster: From the Desert to the Dream," May 6, 2011, https://www.isc.hbs.edu/Documents/resources/courses/moc-course-at-harvard/pdf/student-projects/UAE_(Dubai)_Tourism_2011.pdf, accessed January 24, 2022.

158 Ben Simpfendorfer, *The Rise of the New East: Business Strategies for Success in a World of Increasing Complexity* (New York: Palgrave McMillan, 2014), 135.

159 Guilherme Lohmann, Sascha Albers, Benjamin Koch, and Kathryn Pavlovich, "From hub to tourist destination – An explorative study of Singapore and Dubai's aviation-based transformation," *Journal of Air Transport*, https://doi.org/10.1016/j.jairtraman.2008.07.004, https://research-repository.griffith.edu.au/bitstream/handle/10072/56170/89415_1.pdf?sequence=1, accessed January 24, 2022.

airport offered its own over-the-top theatrics with car giveaways, a hotel inside the airport, and a swimming pool.

Qantas tinkered with shifting its Kangaroo Route hub between Singapore and Dubai after an Australian and Singaporean free trade pact and stock market merger soured.[160] Emirates and Qantas signed a five-year alliance which tarnished the lustre of historical ties on the Kangaroo Route. Political decisions had practical consequences for Qantas passengers hubbing in Dubai, where they could be "left waiting for a connection in Dubai for up to 10 hours."[161] These inconveniences may have contributed to Qantas's decision not to renew its alliance with Dubai's Emirates Airlines. Journalists gushed at the return to Changi, "a thing of rare beauty, an airport that visitors can actually enjoy, a space that seems designed with its patrons in mind."[162] And yet, a stopover's position in the hierarchy of airports was never secure. Whispers about Project Sunrise, promising non-stop service between Sydney and London on new ultra-efficient Airbus A350-1000 or Boeing 787–9 planes, as well as Perth to London non-stops, made for an uncertain future for Singapore's entrepot status.[163]

Plans for a formidable Indian hub to challenge Dubai and Singapore limped into the twenty-first century. In a *Hindustan Times* article of 2003, a handful of Indian executives panned India's putative hubs. One interviewee pointed to the merits of competing hubs: "'We can't compare our airports to Dubai, Paris or London. The shopping here is nothing. There the display is amazing. In Heathrow, cross immigration and there are beautiful shops and boutiques, which are so inviting. 'You can buy the world,'" he noted. Another piled on the execrable state of Indian amenities: "'Our airports are public urinals.'" More pointedly, "'Staff is sloppy. Signages don't exist. The government is doing nothing.'"[164]

Several factors prevented Indian authorities from making nimble adjustments to attract long-haul flights. Resources for an expanded terminal at the existing international airport were not forthcoming. A thriving domestic market also divided

160 Paul Sheehan, "Singapore Left at Altar as Qantas Bounds Off," *The Sydney Morning Herald*, September 10, 2012, Factiva.

161 Michael Gebicki and Robert Upe, "Qantas Travellers Could be Left Waiting in the Wings," *The Sydney Morning Herald*, February 23, 2013, Factiva.

162 Ben Groundwater, "Singapore Fling Time Again," *The Sydney Morning Herald*, February 24, 2018, Factiva. Also see Patrick Hatch, "Qantas plans 'affect alliance,'" *The Sydney Morning Herald*, January 31, 2018, Factiva.

163 On the Perth to London flights, see Michael Gebicki, "A New Long Haul of Fame," *The Sydney Morning Herald*, November 11, 2017, Factiva. On Project Sunrise, see Greg Waldron, "Strategy: Qantas to revisit 'Project Sunrise' at end 2021: Alan Joyce," https://www.flightglobal.com/strategy/qantas-to-revisit-project-sunrise-at-end-2021-alan-joyce/141936.article, accessed January 29, 2022.

164 Nandini R. Iyer, "On the Upgrade Runway," *Hindustan Times*, September 21, 2003, Factiva.

attention between models catering to international or Indian passengers. A gleaming domestic terminal had been completed in Mumbai in the early 2000s, but its distance from the international airport limited any synergies that might have attracted transit passengers. Landing taxes per passenger were also "60% costlier than competing hubs like Dubai, Singapore and Kuala Lumpur and [hurt] India's competitiveness."[165] As a result, one Indian aviation watcher lamented:

> much international traffic into and out of India goes via two major Asian hubs, located at almost ideal transit points – Dubai and Singapore. Both have superb geographic locations and offer very attractive charges to airlines – apart from unbeatable prices to passengers. So great are the transit attractions of these two airports that passengers often prefer to fly by airlines that use them as transit hubs.[166]

Officials at Vistara Airlines, a joint venture between the Tata Corporation and Singapore Airlines, reached similar conclusions, noting that, "Increasingly, Indian passengers travelling abroad are connecting via hubs outside of India leading to loss of revenue, employment and output for the country. As a result Indian airports remain relatively small compared with other global hubs."[167] More specifically, the study noted, "Excluding Delhi and Mumbai which act as gateways to the country, the other airports in the country are lagging in terms of connectivity. Even Delhi and Mumbai compared to global counterparts are not as competitive and this is largely the failure of policy that enables a strong hub carrier."[168] Thus, Singapore and Dubai drew off Indian passengers from national airports not only because of price, but also "transit attractions," among them the highly anticipated Jewel.

To the north, European, Australian, and Western journalists reprised the role of Marco Polo in their encounters with China's sparkling mainland hubs. Australian Joe Parkes wrote, "Cities across Asia are competing to establish themselves as major air transport hubs, building airports of mammoth proportions, capable of handling tens of millions of passengers a year and all designed to lure new business away from their competitors."[169] One such city was Shanghai, newly awakened by sweeping economic reforms.

---

165 Government of India, Ministry of Civil Aviation, *Report of Working Group on Civil Aviation for Formulation of Twelfth Five Year Plan (2012–1017)*, 27. https://www.civilaviation.gov.in/sites/default/files/moca_001320.pdf, accessed June 8, 2022.
166 Hormuz P. Mama, *Second Airport: What Mumbai Must Learn from International Experience* (Mumbai: Observer Research Foundation Mumbai, 2010), 12.
167 Center for Asia Pacific Aviation, "Spectrum Report," n.d., 52, https://www.airvistara.com/content/dam/airvistara/global/english/common/documents/Spectrum_Report.pdf, accessed June 8, 2022.
168 Ibid., 51.
169 Joe Parkes, "Blue Skies are New Skies, and It's Good," *The Australian*, November 1, 1999, Factiva.

Although not a logical competitor with the Southeast Asian and West Asian hubs of Dubai and Singapore, Shanghai's new Pudong International Airport raised the profile of emerging competition between East Asian cities, including Hong Kong. Shanghai's Longhua Airport served as *the* Chinese aviation hub between 1922 and 1949, home to the Chinese National Aviation Corporation among others.[170] Subsequently, Hongqiao Airport served southern China well during the People's Republic era. Like Dubai, it began as a humble airstrip, handling flights between Shanghai and both Beijing and Nanjing. In the 1960s, Air France initiated international flights at the airport in 1966.[171] A new airport, initiated at the behest of Deng Xiaoping, placed Shanghai firmly at the top of the PRC's urban economic internationalisation. Pudong International Airport took flight in earnest in 1995, with the announcement that it would be placed in the new economic district across the river from the Bund. Under the broader initiative, the *Wall Street Journal* rather dramatically announced, "Officials have transformed a swath of farmland almost the size of Singapore into a sprawling construction site in a mere six years."[172] The new airport—designed by French architect Paul Andreu—connected the world to China's new financial showcase.[173]

Paris Airport Company completed the airport design and construction.[174] The Japanese government provided development loans. Among the objectives for the ambitious project was the pressing need to "[build] the city into one of the world's economic, financial and trading centres."[175] The fifteen year target was to build an airport capable of handling between 80 and 100 million passengers, a goal that placed it squarely among the busiest hubs in the world.[176] Not lost on Chinese officials was the zero-sum imperative to "secure its place as the premier commercial and financial center of the Far East – especially in competition with Hong Kong after its reversion to Chinese control next year."[177]

---

170 See Wikipedia, "Shanghai Longhua Airport," https://en.wikipedia.org/wiki/Shanghai_Longhua_Airport, accessed December 5, 2022. For a complete history of its primary carrier, see William M. Leary, Jr., *The Dragon's Wings: The China National Aviation Corporation and the Development of Commercial Aviation in China* (Athens: University of Georgia Press, 1976).

171 *Shanghai Daily*, "Hongqiao Airport Celebrating 100 Years of History," July 19, 2021, Factiva.

172 Joseph Kahn, "Pet Project: Beijing Has $36 Billion Plan to Make Pudong Showcase," *Wall Street Journal*, Asian Edition, August 26, 1996, Factiva.

173 See Philip Jodidio, *Paul Andreu: Architect*, 140–147.

174 Chen Qide, "Airlines Battle for Foothold in Pudong," *Shanghai Star*, August 20, 1996, Factiva.

175 Chen, "Construction of the Pudong International Airport Officially Began on Wednesday," *Shanghai Star*, June 23, 1995, Factiva.

176 Ibid.

177 Geoffrey Murray, "Japan Funds Sought to Build Shanghai's New Airport," *Kyodo News*, September 20, 1996, Factiva.

The airport also served as a launching pad for China Eastern Airlines. Two years after construction began on Pudong International, the airline was listed on the Hong Kong and New York Stock Exchanges. Funds were funneled into a maintenance and storage facility at the new airport.[178] Pudong (PVG) opened to quiet fanfare in September 1999.[179] The airport offered a foothold for European duty-free conglomerates to launch mainland Chinese aspirations. The contract "was a particular coup because it represented the first time that the mainland authorities had allowed the private sector a key role in airport retail services."[180] The third leg of the hubbing stool was not too far in the distance, with "Luxury hotels such as Hyatt and Shangri-La . . . sprouting up to serve business travelers."[181]

In the comparative context of competing Indian and Chinese economies, Chris Leahy cast shadows on India's struggle to embrace the hubbing concept by highlighting Shanghai's head start, writing,

> On arrival at Pudong International Airport, visitors are whisked efficiently through its shiny concourse and off to one of the city's many five-star hotels via a brand new highway. It is hard not to be impressed when you check into the Grand Hyatt Shanghai, billed as the world's highest hotel. It is located on floors 53 to 87 of the 99 storey Jin Mao Tower in Pudong, built on land that less than a decade ago was almost all paddy fields.

Leahy followed this with an unflattering picture of Mumbai: "Housed in a decades-old building with peeling and paintwork and administered by disheveled and apologetic-looking officials, the Chhatrapati Shivaji International Airport [(formerly the Santa Cruz International Airport)] is as difficult to navigate as its name is for westerners to pronounce." Once out of the airport, "the next ordeal is the potholed and traffic-swamped road system to downtown Mumbai. It is difficult to escape the impression that not much has changed since the British left almost 60 years ago."[182]

East Asian airports farther north also took note. "[Incheon International Airport] will face tough competition from fellow international airports in other

---

178 *Financial Times*, "China East Goes Travelling," January 23, 1997, Factiva.

179 Bill Savadove, "Shanghai Opens New Airport, Seeks Economic Boom," *Reuters*, September 16, 1999, Factiva.

180 Sheei Kohli, "BAA Wins Pudong Contract," *South China Morning Post*, August 29, 1999, Factiva.

181 Diane Brady, "Shanghai's New Cultural Gems May Revive its Glorious Past," *The Wall Street Journal*, Europe Edition, September 4, 1998, Factiva. The new hotels included a Hyatt Regency property that would be built into the Skidmore, Owings and Merrill Jin Mao Tower, an iconic nod to Chinese architecture, located in the locus of Pudong's power along the Huangpu River. See *Lloyd's List*, "The Best of East Meets West," August 7, 1999, Factiva.

182 Chris Leahy, "China and India: Foes or Friends," *Euromoney.com*, September 1, 2004, Factiva.

Asian countries, including Chek Lap Kok International Airport in Hong Kong, Kansai International Airport in Osaka, Japan, and Pudong International Airport in Shanghai, China," observed journalist Lee Joo-hee in an article for the *Korea Herald* in September of 2002 – only three years after the opening of Pudong.[183] Airports were not enough in this competitive equation. Their airlines gave them global reach. Ben Sandiland wrote in *Financial Review* that "it's as much Changi Airport versus Chek Lap Kok versus Pudong (Shanghai) as it is Singapore International Airlines versus Cathay Pacific versus China Eastern."[184]

Pudong grew faster than any of the surrounding East Asian hubs during the first two decades of the twenty-first century. Chinese officials announced a second terminal in 2008, which, in the words of a local newspaper, "aim[ed] to make the city a leading global aviation hub by 2010."[185] Wu Nianzu, Shanghai Airport Authority chairman at the time, was more direct: "The target is to make the airport the key hub of Asia's air traffic."[186] Competition, rather than synergies with nearby hub Hong Kong, may have had something to do with the urgency to increase Pudong's capacity. As one journalist noted years earlier near the hand-over to China:

> The bustling British-run territory of Hong Kong, on the approach of its return to Chinese sovereignty, is finding it no longer has a monopoly on connections, confidence and construction. Shanghai, more vibrant than at any time since the 1920s, when it was known as the 'Paris of the East,' is trying to regain center stage in finance, culture and style – and to challenged Hong Kong for its pending title of China's premier city.[187]

Space-aged Maglev transit from airport to city was not the only purpose of the new addition to Pudong International Airport. Twenty-thousand square meters of retail space graced the walkways of the new structure.[188]

Eleven years later, Shanghai officials announced the completion of the world's largest satellite terminal at Pudong, which raised capacity to an astounding 80 million passengers a year. With a third terminal anticipated to bring the total to 130 million passengers, Shanghai, like its skyscrapers in Pudong, entered rarified air. Of the existing capacity at Pudong and the older airport, Hongqiao (across

---

183 Lee Joo-hee, "Incheon International Airport to Become Hub for N.E. Asian Air Traffic," *The Korea Herald*, May 24, 2002, Factiva.

184 Ben Sandiland, "Battle Warms Up for Asia's Crowded Skies," *Financial Review*, Australian Edition, June 17, 2004, Factiva.

185 Lilian Zhang, "Shanghai's Pudong Airport to Open Second Terminal Today," *South China Morning Post*, March 26, 2008, Factiva.

186 Ibid.

187 Steven Mufson, "One China, 2 Cities: Shanghai Seeks No. 1; Hong Kong Has Booming Rival on the Yangtze," *The Washington Post*, June 24, 1997, A11, Factiva.

188 Zhang.

town), *The Shanghai Daily* noted: "Together they handled over 117 million passengers in 2018, a 5.2 percent increase, making it the world's fifth-busiest air hub after London, New York, Tokyo, and Atlanta."[189] The new satellite terminal's retail space dwarfed that of Terminal Two at 28,000 feet of the estimated 620,000 square meters of space covered by the new annexes.[190] This was of vital importance, noted authorities, since "many luxury brand stores at airports saw faster sales growth than at department stores."[191]

At least one aspirant for Asian hubbing supremacy, Kuala Lumpur, made Shanghai-like provisions for bells and whistles, but neglected logistic essentials. Expansive Islamic-inspired architecture that would be complemented with "a Formula-One racing circuit, theme park, hiking trails, golf course and a sprawling shopping center" sat some forty miles from the city center of Malaysia's commercial capital with inadequate planning for mass transit to and from the airport. As one wag noted, "'The trip to the airport will now take longer than most of my flights.'"[192] From a long-term planning perspective, however, the biggest hurdle to global hubbing status materialised in 1972 when the governments of Malaysia and Singapore divided its commercial aircraft fleet and routes between themselves; the former managing domestic routes and the latter, headlined by Singapore Airlines, focusing on international routes. As the *South China Morning Post* incisively noted four years after KLIA's opening:

> Geographically within Asia, there is . . . little separating the two. . . . But what sets the pair apart is that Changi has already built-up economies of scale as a hub and service centre for international airlines. Furthermore, it has a critical mass of passenger traffic through the heavy presence of Singapore Airlines, giving new entrants enough scope of possible passenger transfer combinations.[193]

---

189 *Shanghai Daily*, "World's Largest Satellite Terminal to Open at Pudong Airport Next Week," September 10, 2019, Factiva.

190 Ibid.

191 Zhu Wenqian, "Retail Biz Breathes Life into Airport Revenues," *China Daily* (Hong Kong Edition), January 1, 2022, Factiva. Max Berlinger wrote, "It's no secret that brick-and-mortar shops have struggled of late as U.S. consumers buy more and more goods online. But sales at airport shops have offered up a surprising and welcome counterpoint to that trend. According to Bain & Company, transactions alongside landing strips have increased by 7 percent over the past two years, as opposed to a 4 percent drop at department stores." See "How Flight Delays are Turning Airports into Luxury Malls," *Robb Report*, March 27, 2019, accessed November 19, 2022.

192 Diane Brady, "Bumpy Takeoff? Kuala Lumpur's Airport Garners Mixed Reviews," *Asian Wall Street Journal*, June 29, 1998, Factiva.

193 *South China Morning Post*, "Story of Kuala Lumpur Airport Offers Valuable Lesson," June 18, 2002, Factiva.

The sheer cost and likelihood of replicating, not to mention overtaking, what Singapore had built in the intervening thirty years, courted disappointment. Ultimately, Singapore's prime position as Kangaroo Route entrepot since 1939 overshadowed Kuala Lumpur's regional ambitions.

There were, however, more immediate missteps that handicapped Malaysia's entry to the hubbing wars. Distance to the city center may have diminished enthusiasm for tourists looking for a tourist destination in Southeast Asia, but a shrinking number of international carriers and persistently low passenger numbers doomed KLIA's hubbing aspirations.[194] Qantas, British Airways, and Lufthansa, withdrew from KLIA during the first two years of operations—the first two carriers concentrating on upgraded facilities in Singapore.[195] The intangible impact of Malaysian Airlines twin flight disasters in the early 2000s, one the disappearance of a jumbo jet somewhere over the Indian Ocean, and the other an explosion of a widebody over contested territory in Eastern Ukraine making its way between Amsterdam and Kuala Lumpur, did no favors for confidence in flying to or through Malaysia enroute to Europe, Australia, or closer Asian destinations.

Despite the Malaysian government's dogged persistence to make KLIA into an Asian hub on the order of Hong Kong International Airport, Changi International, or even the problem-plagued new airport in Bangkok, efforts fell short.[196] The solution, it seemed, was to exploit a niche that the Malaysians had pioneering: hubbing for Asian low-cost carriers (the topic of Chapter 5), including Kuala Lumpur-based AirAsia. Initially, plans were made to pull the defunct Subang airport out of mothballs as a low-cost terminal which enjoyed the advantage of easier access to Kuala

---

194 KLIA surfaced as the "sixth most favoured" airport in IATA's *Global Airport Monitor* in 2000. This only ranked "second best Asian airport behind Singapore's Changi," something that would have rankled Malaysian aviation officials. See Asrani Rustam, "Long Flight Ahead for the KLIA," *New Straits Times*, May 23, 2000, Factiva.

195 See Eddie Toh and Neil Behrmann, "BA Move to Scrap Route Seen Hitting KL Air Hub Ambitions," *Business Times*, October 13, 2000, Factiva. Also see *South China Morning Post*, "Story of Kuala Lumpur Airport Offers Valuable Lesson," June 18, 2002, Factiva.

196 Despite facilities for multiple Airbus A-380 planes, as well as the addition of Royal Nepal Airlines, Air China, Philippine Airlines, China Eastern, and Air Kazakhstan, the airport still lagged far behind other regional airports for hubbing status. See Willian Dennis, "KLIA Plans Expansion," *Aviation Week's Airports*, January 13, 2004, Factiva. As late as 2012, Malaysia Airports Holding still touted KLIA's strategic hubbing capacity, but this time as a hybrid entrepot for legacy carriers and low-cost carriers. See Kamarul Azhar, "KLIA Aviation Hub of Southeast Asia?" *The Edge Financial Daily*, March 26, 2012, Factiva.

Lumpur.[197] Subsequently, officials opted to expand KLIA to include a terminal dedicated to low-cost-carriers. Malaysia found success where it had innovated.[198]

Finally, Bangkok entered the hubbing wars in the early 2000s with the announcement of a new international airport to be built over a swamp some thirty kilometers from the city. The Thai government began planning a replacement airport for the heavily transited Don Mueang International Airport as early as 1973. Even after three decades of delays, the likelihood of dual international airports (given the elevated projections of travelers transiting Bangkok to Chang Mai, Phuket, or international destinations farther afield) threatened the hubbing ambitions of Bangkok. United Airlines' Warren Garrett implored Thai authorities to close Don Mueang and focus on the new facility. "The twin airport operation might prompt some airlines to pull out of Thailand in the future," Garrett observed, "because of doubled operating costs and low yields from cheap air tickets."[199] Ultimately, this registered with Thai authorities who aspired to vie with Singapore and Hong Kong for Southeast Asian hub supremacy. Garrett noted that "two airports would affect the country's ambition to become an aviation hub . . . [because] if costs at Suvarnabhumi were too high, carriers would go elsewhere, such as Kuala Lumpur or Singapore."[200]

While the Asian financial crisis of 1997, recurrent political instability,[201] and construction setbacks dogged the inauguration of Bangkok's space-age-like Suvarnabhumi International Airport(rescheduled from 2004 to 2006), "higher operating costs of the new airport and safety concerns over cracked runways at the new

---

197 John Burton, "Subang Set to be Low-Cost Asian Air Hub," *Financial Times*, June 23, 2004, Factiva.

198 Carolyn Hong, "KL Fast-forwards Opening of No-Frills Terminal," *The Straits Times*, June 7, 2005, Factiva.

199 Charoen Kittikanya, "Single Airport, Two-runway Capacity Backed by Airlines," *Bangkok Post*, August 22, 2002, Factiva.

200 Ibid.

201 Infamously, eight thousand protesters occupied Suvarnabhumi Airport's terminal and control tower on November 25, 2008. Three thousand passengers had to be evacuated from the airport the following day with "four explosions . . . early Wednesday morning near the Suvarnabhumi International Airport and Don Mueang domestic airport, injuring at least 12 people." See *Xinhua News Agency*, "Bangkok's Airport Paralyzed after Protestors' Intrusion, Big Loss Expected," November 26, 2008, Factiva. The plight of the passengers is addressed in Newley Purnell's "Thousand Stranded as Protestors Seize Bangkok Airport," *Agence France Presse*, November 26, 2008. As Purnell noted, "Angry travellers who spent the night sleeping on baggage carousels and at check-in desks complained that they had nothing to eat or drink since the protesters burst into the two-year-old terminal."

airport caused many to seek a return to Don Mueang."[202] As one proponent of the old airport noted:

> Everything about Don Muang was wrong. The ceilings were low and claustrophobic. The colour scheme was a combination of dirt-brown/dirt-tan offset by, well, dirt . . . Despite all this, it actually worked. You'd check in, walk into customs, catch a plane and you're out there . . . Suvarnabhumi is terrible in so many ways . . . [cracks] in the runway, control tower blackouts, bad parking, slow baggage and the only airport in the world that proudly puts up signs saying '800m' from arrival gate to the baggage carousel.[203]

A more dispassionate critic opined, "By giving them Don Mueang airport, which can handle an annual flow of 36.5 million passengers if all terminals are used, much of the pressure will be taken off Suvarnabhumi. That will give it a breathing space until the second phase is completed in 2016."[204]

The government relegated all low-cost carriers to the old airport, freeing up capacity for international travelers at the new facility. Bangkok's two airport shuffle made it a formidable regional hub, but nothing close to the global reach of its competitors. "Many of us who have been to the world's best airports—like Incheon in South Korea, Chek Lap Kok in Hong Kong, or Changi in Singapore—know that ours cannot compare," wrote Pawit Mahasarinand for the *Thai News Service*. "They are far superior in terms of signage, clean and comfortable toilets for everyone (not just elite travellers), free wireless internet, sufficient number of sofas and relaxing chairs, and reasonably priced food and drinks, and the list goes on. In those airports, waiting and layover times seem shorter given the better amenities."[205]

Ultimately, Singapore's *coup de grace* in the hubbing war – The Jewel – remained under wraps during its design and construction, which lasted from 2013 to 2018. Changi Airport Group tapped out the services of Moshe Safdie, who also designed the iconic Marina Bay Sands Hotel, to conceive the Jewel. Safdie embedded the world's largest indoor waterfall and a terraced garden within the shopping and gastronomical destination for both citizens and transiting passengers. Safdie noted, "According to [the government's] brief, Jewel would have to be a destination and a major attraction . . . Changi's leadership recognized that to

---

202 "Suvarnabhumi Airport," *Wikipedia*, accessed November 18, 2022.

203 Andrew Biggs, "Terminal Love for the Airport Left Behind." *Bangkok Post*, September 9, 2012, Factiva.

204 *Bangkok Post*, "Airport U-turn Mixed Blessing," June 23, 2012, Factiva.

205 Pawit Mahasarinand, "Getting Higher Ranking for Suvarnbhumi Airport a Costly Exercise in Futility," *Thai News Service*, November 23, 2009, Factiva.

make Jewel the destination it had hoped for, it had to attract passengers to choose Singapore as its endpoint over competitive airports in the region."[206]

Consolidation of a Singaporean, as opposed to European or U.S.-centric aviation-led tourism development model, registered on several levels. True, SIA and the Changi group used jets designed and developed in Seattle and Toulouse, as well as the blandishments of Western consumerism to cater to its transit passengers, but they did so in a way that became associated with the city itself.[207] First, SIA and Changi's financial windfalls represented a state-led model that questioned the low-return model of tourism development based almost exclusively on hotel development. Second, the aesthetics of consumerism were central to attracting transit passengers for relaxation as well as retail therapy. Given the ardor, particularly of Asian shoppers, for "top" brands, there was no contradiction between the presence of exclusive brands in Changi's Terminal 4, largely home to low-cost-carriers. These touches, as well as the highest levels of personal service, were crucial to presenting the best that Singapore had to offer to its' global clientele.

Conclusion

In 1959 Finance Minister Goh Keng Swee established a fiscally sound approach to tourism development by opting for transit-based policies centered on infrastructure development. Such a strategy conformed to the PAP's efforts to bring the economy to life through state-owned entities. Whilst tourism and travel nominally constituted "services," which became a focal point of the period after 1984, the initial thrust of the infrastructure as tourism strategy involved keeping airport facilities below capacity and building a fleet of planes for SIA consistent with a long haul, entrepot specific identity. In terms of travel as tourism, it is noteworthy that while Goh Keng Swee supported Indonesian applications for grants to develop Bali, the Singaporean government did not seek aid from agencies such as World Bank or consultations by United Nations Educational Scientific and Cul-

---

206 Moshe Safdie, "The Design Concept," in *Jewel Changi Airport*, Sam Lubell, ed. (Melbourne: The Images Publishing Group, 2020), 26; also see Stephanie Rosenbloom, "My 27-Hour Vacation in Singapore's Changi Airport," *New York Times*, December 2, 2019, https://www.nytimes.com/2019/12/02/travel/Singapore-Changi-Airport.html?searchResultPosition=1, accessed June 15, 2021.
207 As Engseng Ho notes in his essay, "While many Asian societies today have indeed been shaped by the modern West and reflect it, especially urban ones, there were and are many mobile societies that sustain relations across Asia beyond and before globalization's reach" (908). A vibrant consumer culture that encompasses Asian empires going back centuries would be as indicative of this as anything and resonate temporally with Asian consumerism today.

tural Organization for tourism development. Such projects rarely returned the promised gains from import-substitution tourism strategies. Instead, Goh's assessment of Singapore as primarily the host of transit passengers anticipated the PAP's strategy for catering to the island-nation's strength as a primary node in global transportation through upgrading civil aviation facilities.

This "go-it-alone," "bootstrap" method was deceptive, however. Whilst Singaporean officials funded travel infrastructure and tourism from state coffers, it was not done without significant assistance from the recently departed colonisers, the British. With the old Changi Airfield and millions of dollars to retrain workers from the shuttered British airbase, Singapore enjoyed the windfalls of decolonisation that did not redound to fellow new nations after World War II, namely in Africa. It would fall to those nations, including Kenya, which we noted in passing in the run up to the construction of Changi, to finance airport expansion with loans from the World Bank.

As noted in the introduction, the strategy of travel as tourism took place in two phases. From 1937–1980, Singaporean administrators emphasized destination network growth through promoting Kallang and later Paya Lebar airfields as stop-over points between East Asia and Southeast Asia as well as between Europe and the East. This successful strategy led to the need for a new airport in the late 1970s, a period in global history when an energy crisis ground most economic expansion to a halt. Second, with the development of Changi Airport as an expansion-based enterprise, Singaporean officials shifted from simply touting their nation state as a stopover and aspired to global prominence through increasingly sophisticated strategies to maximise traveler satisfaction at Changi.

The strategy of aviation development as tourism development, in tandem with the ascent of a hometown airline, SIA, gave maximum control to the city-state in building upon the business practices and existing assets of entities which the British had built up between 1819 and 1959. This consolidation of colonial resources played itself out in other post-colonial city states, chief among them Dubai, a tourist destination not above admitting that it took a page from the Singapore playbook to build a diversified economy. And while the aviation as tourism development strategy underscores the global trajectory of Singapore, this study stops short of embracing the revisionist argument that few of the "founding fathers" strategies to economic development were novel.[208] Its founders embraced the export-oriented model of enterprise development in its aviation strategies, a model that other locales throughout the emerging world, including but not

---

208 Barr, 10.

limited to Doha, Qatar; Istanbul, Turkey; and Abu Dhabi, United Arab Emirates have attempted to emulate to varying degrees of success.

This chapter also highlights several scholarly considerations. Most importantly, there was a revival of Southeast Asian driven transit routes throughout Asia and the Middle East traced by airlines including SIA and passing through airports such as Changi. Engseng Ho has signaled the temporal depths of such networks, finding anthropological significance in objects old and new rescued from the past and repurposed for the present age and the future.[209] This corresponds with the idea that patterns of globalisation, indeed, can originate in Asia as well as in the West. This narrative also affirms that such a story transcends the boundaries between East and West in that, in light of Jonathan Bollen's work on aviation routes in the South Pacific and Indian Ocean Basin, Australia as well as Asian destinations found mutual patterns of cultural resonance in the age of mass culture.[210] As Rachel Bok has pointed out, these new networks not only originated in Asia, but retained special meaning there as well.[211]

As the basin-wide efforts of key cities to develop aviation hubs attest, competition for transit passengers remained keen throughout the late twentieth and twenty-first centuries. But the airlines and the airports were only two of the three elements that made the "hubbing for tourists" strategy unique. The third, as Singapore's leaders had worked out, was the presence of world-beating hotels. As the history shows, these luxury hotels worked together with airlines to cater to international travelers almost as soon as commercial flight between Europe and the South Pacific became viable in the late 1920s and early 1930s. Singapore's Raffles Hotel offers the best place to start.

---

209 Ho.
210 Bollen, 5.
211 Bok.

Chapter 3
Asian Hotel Networks in the Age of Aviation

> There are too few [hotels] like this left in the world. Lose as little as you can of it.
> – Legendary Indian hotelier, Rai Bahadur Mohan Singh Oberoi[1]

> Asia's luxury hotels continue to surpass those of any region, especially when it comes to service.
> – American travel writer, Michael Carlton[2]

> The government explained it was not looking to imitate Las Vegas, rather, it wanted a contemporary, bold design representing the Spirit of Singapore.
> – architect Moshe Safdie, on the Marina Bay Sands Hotel project[3]

Introduction

On Thursday, November 2, 1950, columnist Harry Elkan published a feature length article on a recent trip to Singapore, noting that, historically, the legendary Raffles Hotel, was "the first port of call for all tourists, and the resting place for all [British Overseas Airways Corporation] passengers arriving at Tengah or Kallang Airport." In the days before jet service, stopovers were synonymous with accommodation at luxury hotels. Elkan went on to describe the pleasures of Raffles:

> On arrival at the hotel, which has fan shaped palms growing with red stems in the garden, you enter the foyer, follow your Chinese servant to your room. You will pass, on your way glamorous dress shops, jewellers, etc., and you will notice that everything is built to allow air to circulate. Big revolving ceiling fans, and the doors of your room are double swinging doors, with the top and bottom cut off, and only the centre third remaining. No one is going to look under or over the top, so everything is private, and cool. You will find a ballroom, an open-air cabaret, and magnificent dining and social rooms or verandahs.[4]

Elkans' enthusiasm was typical. The Raffles was one of several luxury hotels on the aviation networks linking Australia, Southeast Asia, and Europe from the late 1920s to the late 1950s.

---

1 As quoted in Bachi J. Karkaria, *Dare to Dream: The Life of M.S. Oberoi*, new and revised edition (London: Penguin Portfolio, 1992), 180.
2 *Canberra Times*, "Regent of Sydney Up with Leaders," March 11, 1985.
3 Moshe Safdie, "Rethinking the Public Realm," in *Reaching for the Sky: The Marina Bay Sands Singapore* (ORO Editions, 2013), 28.
4 Harry Ellkans, "Who'll Come A'roaming' With Me?" *Port Lincoln Times*, November 2, 1950, 15.

https://doi.org/10.1515/9783111326641-004

With the advent of jets, which no longer required an obligatory overnight lay-over, and the rise of Western and Asian hotel chains (which were necessary to accommodate the masses disgorged by widebody planes), the lives of the grand colonial hotels languished. Decades later, in 1986, the city-state of Singapore commissioned a study to breathe life into its stagnating tourism sector. Feverish hotel construction in the wake of a 1982 downturn resulted in overcapacity and falling room rates. Tourists also found that Singapore had lost as much as it gained in its rush to exude cosmopolitan chic. "A reason for the shortness of their stay," consultants culled from a government report, "is that the island 'has lost its Oriental mystique and charm' in its zeal to become a modern city.'" Raffles Hotel, however, elicited a promising approach to boosting tourism. As the consultants asserted,

> The Raffles Hotel is quite possibly more famous than Singapore itself. The international marketing potential of its reputation can assist in the overall marketing of Singapore as a visitor destination. In order to realize the full benefit of this potential, the Raffles Hotel must be restored to its former ambience. The restored hotel must be operated in a manner that befits its grand reputation and provides its guests with an experience not available at most hostelries of the world. It should be emphasized that the benefits expected are to the nation as a whole and not just to the owners of the hotel and/or land.[5]

Though in a frightful state of disrepair, this national treasure held any number of advantages for Singapore's tourism development. Luxurious rooms could be re-stored and offered to guests of state. Shops could be curated to satisfy the most discriminating consumers. The Long Bar could be restored to sell Singapore Slings to visitors unable to lodge at the hotel. In essence, leveraging the Raffles to its former glory could revive tourism in a way still appealing to contemporary visi-tors. Raffles was only one of a significant number of colonial hotels that became symbols of national identity in the aftermath of Asian decolonisation.

Two and a half decades later, Singapore's hotel development entered a new phase with completion of the Marina Bay Sands complex. It was the latest in a string of iconic new hotels in places like Dubai and prefigured others such as the InterContinental Wonderland in Shanghai. These hotels were no longer way sta-tions, but destinations that evoked the "Bilbao effect," to use the language of mu-seum development.

These three phases in the life cycle of Singapore hotels illustrate a pattern in the development of hotel networks in twentieth-century Asia. Prior to the jet age, and set amidst the denouement of colonialism, decadent hotels characterised the

---

5 Singapore Tourism Board, *Tourism Development in Singapore: a Report* (Singapore: Singapore Tourism Board, 1986), IV-1.

privileged aviation experience. In the age of decolonisation—with the advent of jets and hotels designed to cater to large numbers of tourists — Asian hotel chains rivalled their Western counterparts, often surpassing them in splendor. Finally, in the early twenty-first century, Asian city states aspired to global status with iconic hotels to match. Aviation entrepots were now destinations with hotels drawing affluent guests from all over the world.

Historiography

Architectural design, post-colonial cultural politics, and heritage studies have distinguished the literature of Southeast Asian hotels. Maurizio Peleggi has written prolifically on tourism development in Southeast Asia. His 2005 study, "Consuming Colonial Nostalgia: The Monumentalisation of Historic Hotels in Urban South-East Asia," for example, examines the subtle distinction between market-driven adaptation of historic hotels for mass tourism and the more museal preservation of properties for historical authenticity. "In the name of development," he concludes, "the hospitality industry is afforded some leeway with regard to national history, whose official representation continues to be found in history museums or the mausoleum of a nation's founding father, where colonialism receives the bad press it deserves."[6]

Joan Henderson has addressed similar concerns in her research on the intersection of heritage status and hotel marketing. Her several articles include "Conserving Colonial Heritage: Raffles Hotel in Singapore" (2001) and "Remembering the Colonial Past: Heritage Hotels and the Legacies of the Sarkies Brothers in Asia" (2018). Henderson concurs with Peleggi in her analysis of Raffles, writing, "The [hotel] illustrates the tensions which exist between conservation and commercialisation of heritage, complicated by the circumstances of colonisation whereby buildings acquire symbolic meanings for both residents and visitors."[7] Her more encompassing article (2018) reaches similar conclusions for the entire suite of Sarkies Brothers' hotels during the late nineteenth and early twentieth centuries, including not only Raffles, but also the Strand (Yangon) and the Eastern and Oriental (Penang). Henderson transitions to postcolonial conclusions, noting that "Guests [of these hotels] are enticed by thoughts of participation in recreations of a privileged lifestyle and hotels are theatres for the enactment of fanta-

---

6 Maurizio Peleggi, "Consuming Colonial Nostalgia: The Monumentalisation of Historic Hotels in urban South-East Asia," *Asia Pacific Viewpoint*, 26:3 (December 2005), 264.
7 Joan C. Henderson, "Conserving Colonial Heritage: Raffles Hotel in Singapore," *International Journal of Heritage Studies*, 7:1 (2001), 7.

sies. In addition, they may become tourist attractions for sightseers seeking to encounter traces of bygone days."[8]

While design plays a secondary role in this chapter, questions of creating intra-Asian networks invoke a different body of literature. The chapter foregrounds hotel and aviation networks. In Peleggi's study, "The Social and Material Life of Colonial Hotels: Comfort Zones as Contact Zones in British Colombo and Singapore, ca. 1870–1930," Colombo and Singapore were two nodes on a seafaring circuit stretching from the Suez Canal to Hong Kong. The Galle Face and Grand Oriental Hotel in Colombo, and Raffles and L'Europe Hotels in Singapore sat along an itinerary to Hong Kong. These institutions featured the technological and cultural amenities familiar to foreign guests. They also served as contact zones between Europeans, local elites, and subaltern employees. Peleggi notes how these establishments presaged "another epochal transition in international travel after the Suez Canal's opening."[9] Extending towards the north and south, "In October 1927 the Royal Dutch Air Service (KLM) operated the maiden flight between Batavia and Singapore; commercial flights between the two cities began eight months later. And in 1929 KLM started the air service between Amsterdam and Java."[10] Frequent refueling created new hotel networks between Europe and Asia.

Taking an exclusively aviation-based approach, Gordon Pirie examines how a predominantly commercial class of air passengers became "incidental" tourists along these routes. As Pirie argues, these layovers:

> [helped] to make tourists of all air travellers irrespective of their trip purpose. Passengers disembarking temporarily at aerodromes could not avoid experiencing the local climate, scenery, buildings and refreshments . . . Imperial [Airways] had its own modest 'station' buildings and facilities in some places, and would recreate comfortable 'little Englands' for its passengers, flying crew and expatriate airfield superintendents.[11]

In these "Little Englands," passengers stayed in Imperial's "rest houses" or hotels of legendary repute along the route from Europe to Southeast Asia. This chapter builds on Pirie's work, illustrating that KLM, Orient Air (forerunner of Air France), and Imperial Airways' respective routes arranged colonial grand hotels into a new

---

8 Joan C. Henderson, "Conserving Colonial Heritage: Raffles Hotel in Singapore," *International Journal of Heritage Studies*, 7:1 (2001), 37.
9 Maurizio Peleggi, "The Social and Material Life of Colonial Hotels: Comfort Zones as Contact Zones in British Colombo and Singapore, ca. 1870–1930," *Journal of Social History*, 46:1 (Fall 2012), 128.
10 Ibid.
11 Gordon Pirie, "Incidental Tourism: British Imperial Air Travel in the 1930s," *Journal of Tourism History*, 1:1 (2009), 56.

network during the early years of intercontinental flight. In one sense, this chapter diverges from Pirie's assessment that Imperial Airways furnished "modest 'station' buildings in some places." Imperial Airways entered the hotelier business at a level of prestige comparable to the grand hotels found elsewhere. As a contemporary newspaper article noted, in a sweeping overview of Imperial's global empire, "the organization has to provide hotel accommodation for passengers at night halts; or, if the halt is at a spot where ordinary hotel accommodation is unavailable, then Imperial Airways operate their own luxuriously equipped rest-houses."[12] This was followed by the establishment of Asian-owned hotel networks in the second half of the twentieth century (which catered to Indo-Pacific businesspeople and foreign tourists—all whisked about on capacious jets), and finally globally aspirational hotel destinations (serviced by jaw-dropping airport hubs) in the twenty-first century.

Politically, this chapter presents a unique set of concerns to those of studies of Western hotels following World War II. While not specifically related to East and Southeast Asia, Annabel Wharton's important study, *Building the Cold War*, established the global intersection of architectural design and Cold War politics in West Asian cities including Istanbul, Cairo, and Teheran. She argued that this new network represented the projection of U.S. soft power in the rapidly decolonising world. Thus, in addition to offering tourists the comforts of modernity, the Hilton network facilitated encounters amongst like-minded natives and foreigners in a modern setting.[13]

To be sure, the Cold War played a key role in Southeast Asian hotel design but was less central to the concerns of Asian hotel owners. As this chapter suggests, American consultants courted governments throughout the region to build hotels catering to the preferences of US tourists. In contrast, Eunice Seng's article, "Temporary Domesticities: The Southeast Asian Hotel as (Re)presentation of modernity, 1968–1973," examines Southeast Asian hotel design within the context of decolonisation. Seng agrees with Wharton that modern architecture defined the form of new hotels in the post-war period. However, Seng argues that decolonisation presented its own set of priorities. Hotel interiors reflected nationalist aspirations. This, Seng contends, was most prolifically illustrated in the work of a Tokyo-educated American interior designer, Dale Keller, whose "orientalised interiors were readily consumed by the economically emancipated [Asian] middle class: many of whom were working in the service industry and also engaged in

---

12 *Hobart Mercury*, "Aviation: Immense British Organisation: Imperial Airways Machines Fly 26,000 Miles Daily," November 25, 1937, 12.

13 See Annabel Wharton, *Building the Cold War: Hilton Hotels International and Modern Architecture* (Chicago: University of Chicago Press, 2001).

office work."[14] Exterior design underscored regional concerns for large scale housing developments. Most importantly, Seng shows that the convertibility of hotels to residences and vice versa further distinguished the form and function of large-scale accommodations in Hong Kong and Southeast Asia.[15]

Seng distinguishes between the functional and stylistic differences in Asian and Western accommodations. This draws Wharton's work back into a discussion of how hotel networks exerted political, cultural, and economic influence in different settings. While Wharton's work suggests that Hilton (and by association other U.S.-based global hotel chains conceived after World War II, including Inter-Continental, Sheraton, Hyatt, and Holiday Inn) aspired to hegemonic, soft power status, the dynamics in Asia were decidedly different. Not only were expectations for hotels less diplomatically fraught, but aspirations for *market* dominance withered because of Asia's sheer scale and the number of players in the hotel market. Indeed, Prasenjit Duara's description of Asian networks generally describes the different waters in which Western hoteliers swam throughout Asia. "Interdependence," he wrote in relation to the personal and public ties in Asia, ". . . is being managed by ad hoc arrangements and specialized transnational institutions with little possibility of large-scale state-like [or market-based] coordination or control. In this sense, region formation [or hospitality market dominance] in Asia is a multipath, uneven, and pluralistic development that is significantly different from European [models]."[16]

Methodology

Writing hotel histories presents unique source limitations. Given the proprietary nature of business records, it is difficult to gather materials beyond advertising materials or corporate histories. As Maurizio Peleggi notes in his study of hotels, "Given the paucity of archival records concerning most colonial hotels . . . their description in contemporary guidebooks and travelogues are very valuable notwithstanding the academic trend to deconstruct such texts as discursive representations."[17] In the same vein, Joan Henderson observes of her work, "Overall findings are based on published materials, assembled from assorted sources as

---

14 Eunice Seng, "Temporary Domesticities: The Southeast Asian hotel as (Re)presentation of Modernity, 1968–1973," *The Journal of Architecture,* 22:6 (2017), 1096.

15 Ibid., 1119.

16 Prasenjit Duara, "Asia Redux: Conceptualizing a Region for our Times," *The Journal of Asian Studies,* 69:4 (November 2010), 981.

17 Peleggi, "The Social and Material Life of Colonial Hotels," 127.

recommended . . ., which include hotel websites, media reporting and academic enquiries. Information was supplemented by fieldwork at hotels which have all been visited by the author, although attempts to secure primary data through interview were unsuccessful."[18]

To be sure, the quality of public histories of hotels has improved significantly. Gretchen Liu's history of Raffles Hotel (1991) and Honkkong and Shanghai Hotels Limited's, *Beyond Hospitality*, take advantage of company records. In an unusually rare case, the manager of the Bangkok Oriental, Giorgio Berlingieri, published *Oriental Album: A collection of pictures and stories of and about the oldest hotel in Thailand*. Finally, Oberoi Hotels commissioned historian Nina Nelson to update an already published history, *Mena House: A Short Story of a Remarkable Hotel*, with a fifth edition. The new edition included coverage of improvements made by the new owners.

Working within these same proprietary constraints, this chapter relies on the same types of published materials noted by Peleggi and Henderson (e.g., corporate histories, advertising materials, and contemporary overviews of the hotel sector), but is supplemented with oral histories, memoirs, and most extensively, primary source accounts of hotels throughout the Indo-Pacific as published in Australian newspapers. Given the distance between Australia and Europe, readers were keen to experience the evolving world of travel even as technologies improved the experience generally. This does not suggest Australians were the only travelers through these networks, but instead that these sources present a systematic way to assess the experience of tourists. As might be imagined, some of these accounts imbibe the racist and culturally hubristic language of an imperial age.

Decadence: Asian Hotels in a Colonial Era (1927–1945)

The opening of the Suez Canal in 1869 accelerated the development of hotel networks in the Indo-Pacific by reducing travel time to Hong Kong and nurturing new cities where passengers might disembark. These networks anticipated stopover points along fledgling air networks in the following century. This extended the economic viability of colonial hotels as well as stimulated the creation of new hotels, particularly in the Persian Gulf, where intercontinental entrepots including Dubai blossomed at the end of the twentieth century.

In 1937 the Strand Film Company and Imperial Airways released a fifteen-minute newsreel entitled, "Air Outpost." It featured a rich description of the air-

---

18 Henderson, "Remembering the Colonial Past," 24.

line's airport-cum-hotel in Sharjah, some twenty-three miles from Dubai. The opening scene distinguishes between the "modern" hotel and the backwardness of the village two miles away. Sharjah's new terminal, the narrator tells us, is one of eighty airports strung along 25,000 miles of British airline routes. The narrator emphasises the remoteness of the airfield, "on the south coast of the Persian Gulf midway between Basra and Karachi, a section of the Indian-Australian Route, which crosses the marshes of the Euphrates delta, the sunbaked deserts of Arabia, and the barren wastes of Baluchistan . . . over which our airlines fly us safely and regularly."[19]

The enclave-like "airport as hotel" figures prominently in the film. The safety of the complex is assured, the narrator reminds us, by its fort-like structure, not to mention an armed Bedouin security force. Food arrives from the village two miles away: "Today as on every day, a merchant from the city is carrying to the airport across the burning desert such fresh food [(in this case a fish dangling awkwardly from a donkey's perch)] as the markets can provide."[20] With the imminent arrival of the Imperial Airways flight at 5:30pm, a uniformed British airfield superintendent beckons his native assistant to duty, admonishing him to prepare rooms and beds for the arriving passengers. The entire native work force springs into action: "Provisions are brought from the store house to supplement local supplies," we are told. "Rooms are gotten ready and beds are made up because Sharjah is a night stop. All through the hot afternoon donkeys bring water from the desert to fill the tank in the courtyard of the fort. Water to provide baths for the passengers and the crew."[21]

It's not just the local subalterns that stir to action. "The airline superintendent will have no time for leisure," the narrator reports, "He must inspect the bedrooms to see that everything is in order, supervise the food, and attend to a multitude of other details."[22] Shortly, the camera turns to passengers alighting from the Imperial Airways plane, attired in dresses or smart suits. In short order, guests are led to their rooms "and handed a card, giving the time of tomorrow's departure, the places at which stops will be made for meals, and all other details of the day's flight."[23]

This vivid account of a passenger's overnight experiences along the Europe to Australia route epitomised the nexus between flight and hotels during the earliest iterations of the Kangaroo Route. While KLM took the lead in launching air

19 Ibid.
20 Ibid.
21 Ibid.
22 Ibid.
23 Ibid.

services in the late 1920s, Imperial Airways and then Air France launched services to their Southeast Asian outposts in the 1930s.[24] KLM eventually reduced flying time in the late 1930s to five or six days. The French route from Paris to Hanoi lasted nearly twice as long, in part because of inclement weather conditions and contentious fly-over agreements with nations along the route. Finally, Imperial Airways consolidated an eight-to-ten-day itinerary, which included a fixed network of hotels and guest houses to match the high expectations of its discriminating passengers. Flight paths were designed to expedite transit between Europe and Australia, The Dutch East Indies, or French Indochina, but also integrated planned stops so that "travellers [might] increase the time they have available for land excursions at chosen points; while by travelling by air over some particular stage they can also avoid or shorten a section of their tour which, by sea, might prove unpleasantly hot or of little interest."[25]

As Gordon Pirie has argued, these scheduled stops made "incidental tourists" of otherwise purposed passengers. What was not left to chance was the quality of accommodations. Publicity materials, including schedules, hyped the luxuriant nature of the stopovers. In Air France's 1938 booklet on Far East travel, copy writers enthused:

> It is only in the different stopping places, that these magic spells are interrupted. It is during these hours that impressions gathered during the flight are memorized and added to the recollection, local sights, seen in haste and curtailed by impending departure. The passenger sleeps in a hotel bed-room, which, in spite of general similarity of decoration, can be remembered either by some outlandish detail, or by a subtle difference in the perfume of the night air.[26]

In retrospect, one passenger noted, "I must say that the accommodation for most of the way was magnificent. In practically every place I had a bed room, sitting room and a bathroom – a suite, which was part of the inclusive fare for the trip."[27] The package tour might also include entertainment, should the itinerary

---

24 Aviation reporter H.E. Baker of the *Queensland Country Life* writes of "Flying Into the Monsoonal Region: Royal Dutch Airways," November 19, 1936. On the travails of French aviation to Indochina see Gregory Charles Seltzer's excellent PhD dissertation, "The Hopes and the Realities of Aviation in French Indochina, 1919–1940," University of Kentucky, 2017, https://uknowledge.uky.edu/history_etds/49/, accessed July 9, 2022.
25 *Rockhampton Morning Bulletin*, "Growing Network of Flying and Shipping Routes: Girdling the Globe by Flying and Ocean Liners," October 26, 1936, 13.
26 "Air France Far East Timetable U.K. edition," March 30, 1938, on Airline Timetable Images, http://timetableimages.com/ttimages/af3803fe.htm, accessed June 24, 2022.
27 *Hobart Mercury*, "Nine Days from London: Former Tasmanian's Impressions of England-Australia Flight," November 3, 1938, 14.

coincide with the presence of dancing troupes or bands touring hotels of Southeast Asia, as happened at the Raffles in Singapore and the Eastern and Oriental Hotel in Penang.[28] Generally, passengers expected the very best, even in the most desolate outposts. As passenger S.M.B. Wansey noted upon landing in Basra, "The airport building contains a hotel. It is not an ordinary hotel with all ancient inconveniences, but a hotel that would do credit to London or New York. Here in Irak [sic], where plumbers are rarer than Frenchmen in the Brown House, each bedroom is provided with a private bathroom complete with running water and green bath, wash basin etc. In the bedroom, fans, mosquito-proof windows, tubular steel furniture, and reading lights attached to each bed prove that 'home was never like this.'"[29] The combination of premiere service in the air and on the ground made for a pleasant experience, given the constraints of commercial aviation at the time. Noted F.E. Lampe, "After about 25 trips across the Equator, . . . this is the most comfortable way of travelling through the tropics. It is impossible to lay too much emphasis on the pleasure, comfort, thrill, enjoyment, and restfulness of this [type of] journey."[30]

As historian Molly Berger has observed, no matter how recent a hotel's novelties, nor superb its creature comforts, many travelers considered hotels incidental to the travel experience.[31] About half of published Australian travel narratives, however, discussed hotels in some detail, perhaps because of their contrast with local conditions. As one anonymous Australian *Women's Weekly* correspondent noted of Baghdad: "by night [it is] sinisterly glamorous, and not bearably smelly. Good accommodation, with bath at the Maude Hotel, and a really excellent evening meal, [was] surprising at Bagdad [sic]."[32] Philip S. Rudder, writing about his Singapore to London flight, effused about the starting point for his trip at "Raffles, [which was] romantic and old fashioned, probably the best known hotel in the world." The trip from the airport to the hotel could be as exhilarating as the flight itself, noted Rudder, who wrote of his landing in Bangkok: "The aerodrome . . . is very large and has numerous hangers . . . We were taken to the city in a Diesel train. Bangkok station is a fine building and reminds one more of the Gare du

---

28 *Philippines Tribune*, "World Famous Dancers at Manila Hotel," June 3, 1936, 6.

29 S.M.B. Wansey, "Across the Mediterranean: Athens to Basra in a Day," *Newcastle Morning Herald and Miners' Advocate*, October 8, 1938, 5.

30 F.E. Lampe, "From London to Melbourne by Air," *Melbourne Herald*, July 25, 1936, 30.

31 Kevin J. James, A.K. Sandoval Strausz, Daniel Maudlin, Maurizio Peleggi, Cedric Humair, and Molly W. Berger, "The Hotel in History: Evolving Perspectives," *Journal of Tourism History*, 9: 1 (2017), 106–108.

32 *Australia Women's Weekly*, "Vivid Diary of Flight South: Woman's Thrilling Record of Eleven Days' Dash from England to Australia," February 27, 1937, 2.

Nord, Paris, than what one would expect in Siam." As for their accommodations? "We stayed at the Oriental Hotel that night. It is an excellent place, and very well run." Neither Raffles nor the Oriental, however, could match Kolkata's Great Eastern Hotel, whose distance from the airport only heightened the anticipation for "a magnificent place, by far the best hotel on the route."[33] A year later, Vernon Heath found the Great Eastern in Kolkata "a fine building [with] good, big bedrooms attached." Going in the opposite direction as Rudder on the same trunk line, Heath then found the Raffles to be "very large, with an open air lounge and dancing floor looking out over the front of the building. [With a] splendid bedroom, dressing-room, hot and cold water." Dinner and a short walk preceded a hasty retreat to bed before a 4am departure.[34]

Some discriminating travelers ranked hotels on the Imperial Airways circuit above those of Europe and the United States. The quasi-residential scale of the rooms in India, as well as impeccable levels of service, impressed weary travelers. As one tourist noted following a trip to Delhi and Agra:

> At the hotel we stepped back into the 20th century, but if it had not been for the style of architecture and the dark faces we might have been in Europe, not India, for we needed a fire in the bed room and five blankets on the bed. Indian hotels are more civilised than Australian [hotels]. Even in what are called 'moderate' hotels the furniture is not confined to a bed and a washstand, designed when Queen Victoria was a fresh and blooming maiden, but takes into consideration that those who stay at hotels do more than eat, sleep and wash their faces. In India each room is large, has a private bathroom, a writing desk, easy chairs and, in some cases, the morning paper is pushed under the door in time to be read with the *chota hazri*.[35]

Other travelers attributed the massive size of Indian hotels, as well as the spacious rooms, to efforts to moderate equatorial climes. Size also accommodated world-weary travelers who would rather engage in the interior "contact zone" of the hotel rather than venture outside into Asia's large cities.[36]

Along the Kangaroo Route

Flying out of London's bustling Croydon Airfield, Imperial Airways passengers on their way to Southeast Asia and Australia looked forward to their last night in

---

33 Philip S. Rudder, "A Trip by Air: From Singapore to London," *Sydney Morning Herald,* July 28, 1934, 11.

34 Vernon Heath, "Gateway to the Indies," *Sydney Daily Telegraph,* November 14, 1935, 6.

35 Norbar, "Journey into India: 5, Delhi, the Imperial City," *West Australian,* March 13, 1937, 7.

36 *Mount Alexander Mail,* "Indian Hotel," July 20, 1885, 2.

Europe at the Hotel Grande Bretagne in Athens, Greece. Baghdad awaited them on their second night.[37] F.E. Lampe, flying in July 1936, arrived in Baghdad around 7:45 pm, aghast at the deteriorating conditions around her. This contrasted with the cocktail and "fresh caviar . . from Persia, Tigris fresh fish, cold turkey and ham" at dinner. Her party then proceeded to the verandah (a must at pre-air-conditioned Middle Eastern and Asian hotels), where they enjoyed "German lager, Egyptian coffee, and fresh fruit." Lampe spent only about half an hour in the city before returning to the hotel. For many travelers, the arrival in Baghdad was a rude introduction to "the East from Suez to Singapore . . . [where the] thrill and gusto of the first contact changes to depressing sadness at the pitiful and wretched sights."[38]

The next day, some remove from the cacophony of the Iraqi capital, passengers alighted at Basra, the hotel adjacent to the airport. Foreshadowing Dubai in a later day, planes from the Netherlands, Germany, and Italy filled the tarmac. W.S.B. Wasmsey's high praise, mentioned above, was not an aberration. Peter Gladwin, flying in 1938, extolled the design of the airport, which provided parallel landing spots for dedicated aircraft as well as flying boats. The hotel rated "magnificent."[39] Consisting of forty-five rooms, the airport hotel garnered glowing reviews from passengers including Geoffrey Tebbutt, who contrasted the "10,000 degree" temperature upon disembarking from the plane, with the "air conditioned bedrooms, lawns, a swimming pool, bright lights, iced drinks, a band in a fly-proof enclosure, and nearly everything that can be done to give limp and transitory Europeans respite from the heat." Tebbutt noted the stark difference of the hotel with Orientalist impressions of the East. "The unchanging East?" he retorted. "This patch of modernity in an oasis is as telling a symbol as might be found of what air transport has brought about."[40]

If Basra prefigured the glitzy hotels and buzzing airports of twenty-first century Gulf destinations, Sharjah's fort, the next stop after Basra, put the future United Arab Emirates on the map. In 1933, Imperial Airways abandoned its original route through the Persian Gulf due to an expiring agreement with the government of Persia. Instead of continuing the service with boat planes, Imperial

---

37 A *Central Queensland Herald* correspondent gives a stimulating description of Croydon in "A Day at London's Airport," June 11, 1936, 5.

38 F.E. Lampe, "From London to Melbourne by Air," *Melbourne Herald*, July 25, 1936, 30.

39 Peter Gladwin, "Boats Fly High to Basra Over Deserts of Iraq," *Sydney Daily Telegraph*, June 29, 1938, 4.

40 J. Percival discusses hotel expansion at the Basra airport in "Modern Airports: A Maharajah Builds an Aerodrome: Basra May Rival Singapore," *Brisbane Telegraph*, March 15, 1938, 13. Tebbutt's comments are included in *Sydney Sun*, "Paradoxes in 8-Day Flight," September 3, 1939, 2.

Airways redeployed the route over land, with planes touching down on the Arabian Peninsula. Building the airport and rest house at Sharjah was no small feat.[41] These far-flung locales, whose sole purpose was to refuel planes unable to connect Karachi and Baghdad directly, also illustrated the advantages of air travel in an age when most opted for sea travel. "Every evening you alight at some convenient air-port," one journalist wrote:

> Finding there everything you need for your comfort until the time comes to an end the next morning. It is such nightly breaks in a long journey which rob it of monotony or fatigue. You have time to explore romantic cities. You secure closeup views of far-off lands, besides seeing them from the air. The time-saving of the air-mail is, in fact, accomplished, not in any breathless rush, but in stages which give you ample time to see all there is to [see].[42]

Notable passengers, including a Lady Willingdon, enjoyed the hospitality of the local sheikh, who "entertained with much picturesque ceremony."[43] Security, however, was the order of the day at Sharjah, with marksmen posted to repulse any attempt to harm the passengers inside the fort.[44]

In the process of forging new networks, the aviation age superannuated some sea-going ports of call. India's Tata family constructed a grand hotel in Mumbai, the Taj Mahal, during the first decade of the twentieth century, intending it to be *the* gateway to India. The owners' aspirations may have been even more ambitious. As one in-house history of the Taj notes, the hotel aspired to be "the 'quintessence of imperial amplitude' and thus instantly joined the 'chain of caravanserais which sumptuously punctuated the travels of latter-day British imperialists': Shepherd's [Hotel] in Cairo, Raffles in Singapore, and The Peninsula in Hong Kong."[45] Technology, however, had not made non-stop flights from Sharjah to Mumbai possible. Instead, Karachi emerged as the aviation gateway to the Indian subcontinent in the early 1930s. During the age of flying boats, travel narratives said very little about accommodations at Karachi. With construction of an expanded airport following World War II, however, workers converted military quarters into a luxurious hotel. One tantalising review recounted staying at a "beautifully appointed hotel" where the "marble floors could be appreciated by

---

41 Montague Webb, "By Air Mail to India: The Arabian Coast Route – Experience of a Year," *Perth Sunday Times*, October 22, 1933, 9.
42 *Newcastle Morning Herald and Miner's Advocate*, "Flying Mail: Famous Air Passengers, Kings, Statesmen, Sheiks," September 1, 1933, 19.
43 Ibid.
44 *Brisbane Telegraph*, "Airway Travelling Over Desert and Jungle," December 18, 1937, 20.
45 Taj Mahal Hotels, *The Centenary: Taj, 100 Years of Glory* (Mumbai, 1993), n.p.

walking in bare feet."[46] Forty well-appointed rooms, a cinema, and portraits of Pakistani nationalist heroes also welcomed guests in the post-World War II era.[47]

Jodhpur and Allahabad's new airports and hotels further vindicated the idea that India's, and indeed the Middle East's, traditional leaders were no less attuned to modernity than the jetsetters they hosted. With eye-popping contrasts between the venerable palaces of maharajahs of yesteryear and the new airport at Allahabad, passengers offered enthusiastic reviews for the new facilities. While an Orientalised view of the cityscape sometimes influenced views on hotels generally, the Allahabad rest house held its own, even against the stiff standards of Europe. "The airport hotel," raved J. Percival, "is outstandingly the best between Australia and England. The bedrooms are of the most modern style with fireplaces, electric fans, hot and cold water, and every comfort than can be desired." Luxury appointments remained close at hand as "the dining room, with its old English furniture and stucco work on the walls is a work of art."[48] For another traveler, the immaculate hotel reflected the *maharajah's* "progressive spirit." "A swimming pool and ballroom are outstanding luxuries in any hotel," S.M.B. Wansey crowed, "but in dreary Rajputana they are unique. Each visitor, moreover, occupies a suite-bedroom, dressing-room, and bathroom with plumbing. Flyproof doors and windows, and wooden shutters keep out insects and fierce light. Bare-footed, turbaned Indians, wearing white gowns with coloured cummerbunds, wait at tables. European food, cutlery as clean as it was on leaving Sheffield."[49]

Unlike the modern airport hotel facilities of Basra, Sharjah, Jodhpur and Allahabad, six of the final seven overnight entrepots featured hotels steeped in nineteenth century colonial splendor. Kolkata was the final overnight stay on the Indian subcontinent. Passengers stayed at the oldest known luxury hotel in India, the Great Eastern Hotel. The hotel had established itself as the principal accommodation for discriminating Australians attending the annual Durbar. It subsequently took on new life as a stopover on the aviation circuit. As several travelers noted earlier in the chapter, it was the last word in elegance east of Delhi. World traveler Leo Kieran delivered a dispatch from the hotel, comparing it to a "mu-

---

46 *Mount Gambier Border Watch*, "On the River at Basra, A Port on the Air Mail Route," July 17, 1948, 5.
47 *Queensland Times*, "Karachi: Air Capital," December 19, 1949, 6.
48 J. Percival, "Modern Airports: A Maharajah Builds an Aerodrome: Basra May Rival Singapore," *Brisbane Telegraph*, March 15, 1938, 13.
49 "Flight from Holland: Across the Persian Gulf; The Glamour of the East," *Newcastle Morning Herald and Miners' Advocate*, October 15, 1938, 5.

seum," whose rooms, in accordance with the Asian preference for residential dimensions, "were huge."[50]

Farther east, Bangkok's legendary Oriental Hotel, opened in 1860, was, like the multi-purpose rest houses in the Persian Gulf, "a self-contained world, which . . . combined all the aspects of hospitality: accommodation, catering, laundry, exchange services, telephone operators, sports and social facilities like clubs or regular meeting areas."[51] Curiously, popular histories of the Oriental rarely mention the hotel as a stopover for air passengers. Given the overwhelming attention to visiting dignitaries and literary figures, it is likely that readers were more interested in fashioning cultural "imaginaries."[52] Similarly, former manager and historian, Giorgio Berlingieri gave only a passing nod to "A New Era," which commenced in the 1930s with "The first group of aerial travelers [landing at Bangkok's] Don Mueang." Feted by the local press, for example, Mr. and Mrs. C.H. Day "made their way into town to become the Oriental's first air-borne customers."[53] Perhaps few passengers commented on the Oriental because they spent so much time roaming the city. J.V. Fairbairn reported, "we got to bed only two hours before being called [for the taxi to the airport]." To compensate, however, "It is so restful on the aeroplane all day that we need exercise [during the night] rather than sleep."[54]

Two generations of the Armenian Sarkies family managed the next four stop-over points. While their hotels—The Strand in Yangon, Myanmar; the Eastern and Oriental in Penang, Malaya; Raffles Hotel in Singapore; and, the Oranje Hotel in Surabaya, Dutch East Indies—catered almost exclusively to European guests, readers must exercise caution in attributing imperial identity to the Sarkies family itself. Like the Kadoorie and Sassoon families in Hong Kong and Shanghai, respectively, these hoteliers from Persia and Iraq were as much West Asian or Middle Eastern as they were culturally anglicised. Those ethnic and geographic distinctions should count as much for their familiarity with decadent cultural forms of design and entertainment—consistent with East and Southeast Asian colonial hotel form and function—as their understanding of their clientele's prefer-

---

50 Leo Kieran, "Hop Over Jungles Marks World Trip," *The New York Times*, October 12, 1936, 28 *Proquest Newspaper Database*.

51 Andreas Augustin and Andrew Williamson, *The Oriental Bangkok,* in the series "The Most Famous Hotels in the World" (London: The Most Famous Hotels in the World, Ltd.), 32.

52 See Noel B. Salazar, "Tourism Imaginaries: A Conceptual Approach," *Annals of Tourism Research*, 20:2 (2011), 865–866.

53 Giorgio Berlingieri, *Oriental Album: A collection of pictures and stories of and about the oldest hotel in Thailand* (Bangkok: D.K. Book House, 1970), 32.

54 J.V. Fairbairn.

ences.[55] As hotel historian Andreas Augustin perceptively notes in his study of Yangon's Strand, "Again, like [the case of Raffles Hotel] in Singapore, we are confronted with the irony that this is considered to be another of the 'most British' and 'colonial' hotels of Southeast Asia. In fact it was conceived, built and run by an *Armenian* family firm. Never ever was it British."[56] Augustin made a similar comment during his study of the Strand, which linked the commerce of the old Silk Road to the contemporary age of aviation in the early twentieth century, noting that the Sarkies brothers:

> Developed a new mixture of Asian-European hospitality hitherto unknown in Southeast Asia. Their background as traders and merchants, coupled with their lack of experience in international hospitality, helped them to develop a new view of the actual needs of potential guests, unhampered by prevailing notions. With Persian connections, they brought caviar from the Caspian Sea to Penang, Singapore, and Yangon. For almost five decades, they built, opened and ran hotels which have become legends in their own right.[57]

Passenger expectations ran high at the Sarkies' properties. Aviation breathed new life into some hotels passed over by ocean routes, including the Strand in Yangon. With the advent of flying, "those bay-destinations were perfectly situated to become regional traffic-hubs."[58] While some aviation historians may beg to differ, Andreas Augustin argues that "Yangon became the most important hub on the air route from Europe to Southeast Asia. The British were clever enough to quickly adapt the airport to the needs of the time and so Yangon was looking at a bright future."[59] In any case, Yangon's favourable location extended the life of the

---

55 For more on the role of Armenians and Baghdadi Jews, including their cultural ties to their Middle Eastern provenance, see Margaret Sarkissian, "Armenians in South-East Asia," *Crossroads: An Inter-disciplinary Journal of Southeast Asian Studies*, 1987, 3:2 (1987), 1–33. Sarkissian makes abundantly clear that the Armenians were dispersed during the Safavid period from Persia, fanning out across Southeast Asia. Nadiad H. Wright has written extensively about the small, but influential Armenian community in Singapore, including her book, *The Armenians of Singapore: A Short History* (Penang, Malaysia: Entrepot Publishing, 2015). On the cultural affiliation of East Asia's Baghdadi Jews, see Chiara Betta, "Orientals to Imagined Britons: Baghdadi Jews in Shanghai," *Modern Asian Studies*, October 2003, 37:4, 999–1023. Betta notes that while the Sassoon family, and David Sassoon in particular (owner of several hotels and apartment complexes in pre-World War II Shanghai), aspired to inclusion in the tightly knit British cultural community in the International Settlement, the family remained culturally linked to their ancestral homeland.

56 Andreas Augustin, *The Incredible Tale of the Legendary Strand, The Most Famous Hotel of Rangoon*, in the series "The Most Famous Hotels in the World," (The Most Famous Hotels in the World, 2017), 9.

57 Ibid., 22, emphasis mine.

58 Ibid., 89.

59 Ibid., 94.

Strand Hotel, where "The topics of the tales swapped at the Strand Bar changed from fancy-dress balls and deck tennis to air turbulences over Baghdad and a broken propeller near Kashmir."[60]

Australian travel narratives rarely mention the Strand in Yangon, but L.E. Lampe noted a stay at the Minto Mansions there in 1936.[61] The Minto was well regarded for managing a refrigerated kiosk for passenger refreshment at the Yangon airfield.[62] Singapore's Raffles, in contrast, found its way into nearly every narrative of the era. In 1936, J. V. Fairbairne noted the generous size of the rooms, writing, "My suite in the Raffles Hotel at Singapore was 17 yards square and furnished with very good modern furniture. It gave me such a surprise that I thought there must have been a mistake."[63] The same year, H.E. Baker reported on a night landing, where taxis whisked guests the twenty miles to the Raffles Hotel. There, the author found "a very large, commodious and comfortable place."[64] Not only were the public spaces of the hotel ample, but the entertainment kept air passengers amused into the small hours of the morning. Baker noted, "[Raffles] is centrally located, but what with dance every night till 4 am there sometimes being as many as 1,000 dancers and street traffic all night, it is much too noisy." The swimming pool, nonetheless, compensated for any irritation, as it was "delightfully situated and overlooks the sea."[65]

Paradoxically, Raffles later suffered the same fate as Mumbai's Taj Mahal with advances in aircraft technology and infrastructure. With the shifting of the airport from Kallang to Paya Lebar and then to Changi, more and more modern hotels dotted the Singapore skyline. Furthermore, by the mid-1960s, as Raffles historian Gretchen Liu notes, the hotel "could no longer depend on overnight transit passengers; planes had begun to fly through the night."[66] Politics and wear also took their toll as "the patina of age took on a decidedly negative connotation [and] Asians and western radicals could not separate the hotel from its colonial associations."[67] It would not be until the late 1980s that the city-state deemed it

---

60 Ibid., 91.

61 F.E. Lampe, 30.

62 W.A. Baird, "Log of Air Journey from England: Jodhpur to Kolkata," *Brisbane Telegraph*, September 12, 1935, 15.

63 J.V. Fairbairn, "To London by Air," *Melbourne Argus*, June 13, 1936, 10.

64 H.E. Baker, "To India by Air: In the Clouds Over Java," *Queensland Country Life*, September 17, 1936, 5.

65 Ibid.

66 Liu, 146. Ironically, it was Qantas that "celebrated the inauguration of the Boeing 707 [jet] service with an informal luncheon at Raffles Hotel." See *The Straits Times*, "The Boeing 707 Cuts Old Flying Time by Half on First Flight from Sydney to Singapore," October 28, 1959, 1, https://eresources.nlb.gov.sg/newspapers/, accessed April 14, 2023.

67 Ibid.

politic to acquire the building, restore it for the tourist trade, and push the colonial legend as a state-owned vehicle for national tourism development.

East to Hong Kong and Shanghai

In the Far East, Hong Kong and Shanghai hotels buzzed with the energy of the new century. One of the main players in the hospitality industry was the consolidated Hongkong and Shanghai Hotels, Limited, controlled by the Baghdadi Kadoorie family. With a clutch of hotels in Shanghai, including the Astor House, the Majestic, and the Palace, the company aspired to "[control] the leading hotels in North China as well as . . . in the South . . . becoming allied with the other leading hotel enterprises in this section of the Far East."[68]

In the early 1920s, the HongKong and Shanghai Hotel Company, Ltd. conceived of a luxury hotel near the railway station that would serve passengers *enroute* to China and beyond. Billed as an "itinerant hotel" in the language of the day, its appointments would be nothing if not magnificent. The *Hong Kong Telegraph* announced initial plans for the Peninsula on August 26, 1921, noting that it would:

> be constructed on entirely modern lines and containing a very fine lounge and entrance hall, a modern ballroom and roof garden, suites of rooms and a limited number of suites de lux. Each room will have a self-contained bath room, with modern plumbing installed. The Hotel will also have its own self-contained hospital and operating theatre, and many other features entirely new to the Far East.[69]

Although the hotel opening was delayed several times, including for construction setbacks in 1923 and temporary occupation by the British army in 1927, HongKong and Shanghai Hotel Company officials urged patience. "On completion there is no doubt that they new hotel, which will cost in the region of $3,000,000, will be a long way ahead of anything of its kind in the Far East," one newspaper article assured readers.[70] An update two years later in the *Hong Kong Telegraph* promised a "view from the harbour to be obtained from the upper floor [that] is one

---

68 HongKong and Shanghai Hotels, Limited, *Beyond Hospitality*, 61.

69 *The Hong Kong Telegraph*, "Local Enterprise; The Hongkong Hotel Company's Activities," August 26, 1921, https://mmis.hkpl.gov.hk/old-hk-collection, accessed April 14, 2023 (cited hereafter as OHKC).

70 *The Hong Kong Telegraph*, "Hong Kong Hotel Meeting; Peninsula Hotel Next Year," April 28, 1923, 1, OHKC.

of absorbing interest and delight."[71] Fully aware of the commercial appeal of Hong Kong to its guests, the *Telegraph* continued, "[on] the ground floor . . . will be located a double row of shops served by a lofty and well-lighted arcade through which entrance can be obtained into the Hotel proper."[72]

On December 11, 1928, a band played on the Roof Garden as local dignitaries toured the much-anticipated hotel during its grand opening. Newspapers reported on the three-pronged significance of the new hotel. First, the building evoked admiration as attendees were "delighted with the sumptuous decoration and the modernity of the general equipment."[73] Second, J.H. Taggart, Managing Director of the Hongkong and Shanghai Hotels Company, Ltd., spoke to its significance as a gathering point for passengers in transit. "If I may be permitted to say so," he intoned,

> I think that this edifice may justly be regarded as not only a worthy addition to the amenities of every day life of the residents of this Colony, but also (in view of its location at the gate-way to this great port and the fact that it is superlatively equipped for its particular sphere of activity) as an establishment which may fairly be expected to enhance the popularity of the Colony by affording to transient visitors thereto attractions in the way of comfort and facilities hitherto unattainable here.[74]

Finally, Taggart spoke to the hotel's promise to elevate Kowloon and Hong Kong, along with other recently constructed landmarks, including City Hall, to greater global significance as an entrepot. Ships plied the harbor of Hong Kong not far from "the terminus of what will one day be the through railway line from Europe, and with a fine modern aerodrome [at Kai Tak] now in course of construction – a city with a future far exceeding the possibilities of the cramped surroundings of Victoria."[75]

The Peninsula served not only as an accommodations entrepot in Britain's commercial Asian outpost, but as a cultural gathering place of global significance. Not three weeks after the hotel's opening, the Filipino community gathered on

---

71 *The Hong Kong Telegraph*, "Hotel Company Meeting; Proposed New Share Issue; Peninsula Hotel Ready Next Year," October 20, 1925, 1, OHKC.

72 Ibid. A year after the hotel's grand opening, the *China Mail* later reported, a luxury silk shop managed by Mr. V. Shewaram opened "in the presence of a large gathering which included all the members of the local Parsee community, the Sindi silk merchants of Hong Kong, and many European ladies and men" ("Peninsula Hotel; The Pioneer Silk Store's New Enterprise," January 17, 1929, 1), OHKC.

73 *Hong Kong Telegraph*, "Peninsula Hotel Opening; H.E. on the Future of Kowloon; A Fine Building," December 12, 1928, 2, OHKC.

74 Ibid., OHCK.

75 Ibid, OHKC.

the Roof Garden at "The Pen" to commemorate the death of national patriot Jose Rizal "by holding a concert and dance." Compatriots contributed to the exposition of European songs as "there [was] a considerable amount of musical talent among the local Filipinos."[76] The Rizal memorial epitomised the Peninsula's reputation as the ultimate setting for musical production in the colony. The Peninsula's manager, A. G. Piovanelli, hired a new band in 1932, the Capitalians, to great public acclaim. Brought in after a three-year stretch in Hollywood, the *Hong Kong Telegraph* reported, "many encores were insistently requested by the public who were unanimous in affirming that the new band of the Peninsula is certainly the very best one in the Colony."[77] Other ethnic and cultural communities also held their gala balls at the Peninsula. In 1930, The St. Andrews Ball was held the last Friday in November. "The last clan shield has been put in position, the last piece of tartan has been hung, and the dining and dance rooms are bristling with thistles, escutcheons and mottoes in readiness for the merry-making throng that will invade the Peninsula Hotel tonight," *The Hong Kong Telegraph* reported.[78] Five years later, the Australian and New Zealand Association of Hong Kong held a ball at the hotel, with "Australian Gum Bushes . . . decorating the Rose Room, while the stars of the Southern Cross shone out from the flag which draped the centre lights."[79]

With the advent of regularly scheduled commercial aviation at Kai Tak Airport, the Peninsula transitioned its focus from railway passengers to the arrival of guests by planes. Making a figurative connection between air travel and the properties in 1953, Freda Irving pointed out that "the Peninsula Hotel is the terminal for all airlines touching down in HongKong."[80] Historian Andreas Augustin, who lived in residence at the hotel while writing a short sketch of the hotel's origins, began his book from just that perspective, writing:

> The Peninsula is the largest hotel in the world. It starts at the airport and comes to an end five miles away in the centre of bustling Hong Kong. You come out of the sliding doors at the airport and the hotel magically appears in the shape of a Rolls Royce Silver Shadow. . . . Suddenly the harbour comes in sight. To the right appears a wonderful old building with

76 *The China Mail*, "Filipino Patriot; A Commemoration gathering; At Peninsula Hotel," December 31, 1928, 7, OHKC.

77 *The Hong Kong Telegraph*, "A Gala Night; Evening's Enjoyment at Peninsula Hotel," May 2, 1932, 2, OHKC.

78 *The Hong Kong Telegraph*, "Caledonian Ball Features; Peninsula Hotel Stage Set," November 28, 1930, 1, OHKC.

79 *Hong Kong Daily Press*, "Fancy Dress Ball; Peninsula Hotel Attraction," April 3, 1937, 1, OHKC.

80 *Melbourne Argus*, "Career Woman in Hong Kong," May 7, 1953, 10.

outstretched wings. The Rolls makes a turn in front of the Star Ferry and the driver steers his flying carpet back to that old façade you had just been admiring: The Peninsula.[81]

Figurative imagery aside, the early years of The Peninsula featured prominently in the evolution of aviation and tourism in the city-state. Augustin later records that one entrepreneur "ran his travel agency in the Lobby . . . where Cathay Pacific manager Tommy Box or his opposite number, Walter Nollorth of Philippine Airlines, who both had their offices in the hotel, also met."[82] Pan American Airways also had an office in the Peninsula in the 1960s.[83] By the turn of the twenty-first century, the hotel complemented their Rolls-Royce service to the airport with "two rooftop helipads for its Aerospatiale Squirrel helicopter, with guests arriving and departing through the historic [China] Clipper Club."[84]

Travelers flocked to these hotels during stopovers and dedicated trips to the shopping hub of East Asia.[85] The views, the service, and the finely appointed rooms all garnered fulsome praise. "We stayed at the Peninsula Hotel, a very fine hotel, the view of Hong Kong at night from the top being glorious," one E.S. Corser boasted. "The city is built at the bottom of the hills which stand up proudly behind. Numbers of modern buildings are built on the slopes and right up the very skyline and when these are all lighted at night, together with the varied coloured lights in the city, you can imagine the beautiful effect."[86] Business reports hailed the island as a must-see for tourists, and the HongKong and Shanghai Hotels Company Ltd., sat at the zenith of comfort. Noted one writer, "'Hong Kong is referred to as the Riviera of the East. It is a port of call for round-the-world tourists. The hotel accommodation is excellent. The Hong Kong Hotel, Peninsula Hotel, Repulse Bay Hotel, and the Gloucester Hotel comparing with any leading hotel in Sydney.'"[87]

---

81 Andreas Augustin, *The Peninsula*, in the series, "The Most Famous Hotels in the World Series," 9–11 (1991).

82 Ibid., 71–72.

83 Pan American Airways, *Blue Book of Clipper Travel* (New York: Pan American Airways, 1968), 70, https://digitalcollections.library.miami.edu/digital/collection/asm0341/id/98717/, accessed April 14, 2023.

84 *The Advertiser* (Adelaide), "Flying Airport to Hotel," February 17, 2001, Factiva.

85 Anecdotally, Mr. T.J. Mildren, manager of the Repulse Bay Hotel conjectured during a trip through Australia in 1935, "'You will be surprised to learn that more than 60 percent of the tourists who have visited the Orient during the past year hail from Australia,'" (*Brisbane Telegraph*, "Orient Attracts Australians," November 13, 1935, 6).

86 E.S. Corser, "Trip to the East: Glimpses of Life at Canton; 'Teeming Millions,'" *Maryborough Chronicle, Wide Bay and Burnett Advertiser*, July 3, 1936, 6.

87 *Inverell Times*, "Hong Kong; British Island Colony; Important Possession; 'Riviera of the East,'" January 26, 1940.

Journalists often mentioned the hotel's high-profile automobile fleets linking the airport with hotel accommodation. Not only did the Peninsula maintain a fleet of Rolls Royce's to whisk passengers from the port or landing field to the Kowloon property, but also a host of thirty Studebaker's stood at the ready to carry tourists to the Repulse Bay Hotel.[88] Australian authorities also found the Peninsula to be a boon to the island's development, as it hosted an Empire Fair featuring Australian agricultural products on its ground floors during late May 1933, Attended by a reported 200,000 visitors.[89]

Unlike the colonial properties throughout Southeast Asia, the Hongkong and Shanghai's Hong Kong hotels transitioned more easily to the late twentieth century. Discriminating travel writer Walter Reisender emphatically noted:

> [One] establishment can be relied upon to stay as dependable as ever—the Peninsula Hotel, grand duchess of Hong Kong's fine hotels, which has become a legend among tourists and businessmen. The [Eastern and Oriental] in Penang, Frank Lloyd Wright's Imperial Hotel in Tokyo, Raffles in Singapore, Mount Lavinia in Colombo – all had difficulty coping with change. One hotel which managed the transition is the Peninsula in Hong Kong, a . . . sumptuous and truly grand 210-room hotel in the European manner.[90]

This was largely the result of Horace Kadoorie's thirty-five years (1950–1985) at the helm of the Peninsula.[91] This sparkling reputation was also a result of superior service. As travel writer Alfred Barcover pointed out to his readers in the *Chicago Tribune*, superior attention from the staff was at the heart of the Asian hotel experience. "Service might be dying in some major cities but not here."[92] Another factor was management efforts to keep up with increasingly congested air traffic. "'With the coming of the Jumbo jet,'" An official from the Hong Kong Tourist and Travel Service, noted, "'I don't know where we will put these visitors or how we can. Move them once they clatter down the ramps of the aircraft. Service transportation and hotel accommodations will be our chief problems.'"[93] "The Pen"

---

88 *Adelaide Register*, "Thirty Studebakers Link Hong Kong and Repulse Bay," June 18, 1924, 5.

89 *Melbourne Herald*, "Showboat for Hong Kong? Supplementary Plan Prepared; Big Fair in May," November 21, 1932, 5.

90 Walter Reisender, "The Stately Jewel of Hong Kong," *Australia Jewish Times*, November 26, 1987, 10.

91 For an overview of the Kadoorie family contributions to Hong Kong, including humanitarian efforts, see "Kadoorie: A Family Legend," *Australian Jewish News*, November 4, 1988, 55.

92 Alfred Borcover, "Hong Kong: The Manhattan of the Far East: It's Cosmopolitan, Beautiful, Exciting – and Tops for Shopping," *Chicago Tribune*, September 26, 1982, H1 *ProQuest Newspaper Database*.

93 Leavitt F. Morris, "Making Room for Hong Kong Book: Travel Editor's Diary," *The Christian Science Monitor*, March 18, 1968, 13 *ProQuest Newspaper Database*. Contemporary newspaper accounts not only portrayed Hong Kong as a destination in its own right but advocated what was

stood at the ready. To add emphasis to the linkages between aviation and hotels, architectural historian Eunice Seng notes that "In January 1974, a strategic alliance was made to appoint Cathay Pacific Airways as partner to promote the Peninsula Group's hotels and accept reservations on their behalf in its eighty offices throughout the world."[94]

Some even ventured that the city's hotels surpassed what was on offer in East Asia during the go-go age of the 1930s. Travel writer Horace Sutton opined, "If anything, Hong Kong is more luxurious and indulgent than anything that existed on mainland China even in the days when Shanghai was in part a European city and the Empress Dowager fanned the humid summer days in her splendid lakeside palace outside [Beijing]."[95] The Peninsula kept pace through innovation, challenging newcomers like Hilton International.[96] To this end, the Hongkong and Shanghai Hotels, Limited, did something heritage enthusiasts refused to countenance in Singapore: it built a thirty-story tower adjacent to its historical property.[97] It also continued its Rolls Royce transfer service, a clear nod to the pre-jet days as experienced by early flyers throughout Asia.[98] It upgraded its shopping gallery, Sutton noting that "svelte little corners all over the hotel sparkle with such familiarly named outposts as Hermes of Paris, Ferragamo of Florence, Gucci of Rome and Cartier of Paris (featuring this week a Buddha with fold ear loops)."[99] Finally, it featured curated experiences for mainland tourists.[100]

---

termed in the previous chapter as an "Entrepot for Tourists." Journalist Kermit Holt wrote a decade later, "For the nearly-exhausted travelers who have had the foresight to schedule a three- or four-day stop here at the midpoint or end of a lengthy tour of the Orient, those who have literally staggered through temple after temple, and ruin after ruin, day after day for weeks, Hong Kong is something of a paradise." See Kermit Holt, "Hong Kong and its Islands: 'Rest stop' on Tour of Orient," *Chicago Tribune*, April 3, 1977, C1 *ProQuest Newspaper Database*.

94 Seng, 1118.

95 Horace Sutton, "Travel With Me; Hong Kong: Lap of Luxury sits at Mainland China's Side," *The Baltimore Sun*, January 7, 1979, H4 *ProQuest Newspaper Database*.

96 *Beyond Hospitality*, 123.

97 *Australia Jewish News*, "Peninsula's Luxury 30-storey Tower," June 10, 1994, 22. Also see *Australian Jewish Times*, "New Life for the Old Peninsula," April 8, 1988, 14.

98 Reisender, "The Stately Jewel of Hong Kong," *Australian Jewish Times*, 10, notes that after arriving at Hong Kong's Kai Tak airport "you are on your way [to the Peninsula], sitting in the air-conditioned, leather scented comfort of one of the [hotel's] nine Brewster-green Silver Shadows – the world's largest private fleet of Rolls-Royces."

99 Sutton.

100 See "Cultural Experience at Hong Kong Peninsula," *Australian Jewish News*, December 18, 1994, 4. With this package "guests start their morning with tai chi exercises from a master teacher on the garden terrace. After breakfast there is a tour of the hotel's Chinese kitchen conducted by a leading chef, followed by a hand-on demonstration of making dim sum. Another ac-

If Hong Kong's Peninsula Hotel was the pinnacle of elegance in Asian hotel history, its Shanghai counterparts, both those of the Baghdadi Kadoorie's and their rivals, the Sassoon's, represented the essence of risk and reward in the field. After managing the wildly successful Astor House Hotel located on the Bund, the Kadoories built and managed the Majestic Hotel in Shanghai, site of the wedding reception of Generalissimo Chiang Kai-Shek and his bride Soong Mei-Ling. Victor Sassoon, the *bon vivant* grandson of Mumbai's Baghdadi transplant, David Sassoon, sensed the need not only for what at the time might have been considered "seven star" hotel opulence of the highest order in this Shanghai, but also new residential spaces in rapidly growing Asian metropolises. This was not lost on the Kadoories, whose own clientele saw the city's luxury hotels as "a home away from home . . . a refuge where China was cast into the background."[101]

Sassoon constructed a clutch of apartment buildings in Shanghai before completing the Cathay Hotel in 1929.[102] This new sixteen story skyscraper on the Bund pushed the limits of hotel amenities, decadence in design, and peerless service in a market that had been dominated by the Kadoorie's. Sassoon took the final ruthless step of purchasing and putting out of business the Kadoorie's Majestic, turning it into apartments. A Perth newspaper objectively laid out the scale of Sassoon's new hotel, The Cathay: "215 rooms and suites, each with private bath; restaurants with spring dance floor opening on to roof gardens and terraces; a la carte restaurant on the nineth floor; banqueting and private dining rooms on tenth and eleventh floors; all public rooms equipped with air-cooling ventilating system, allowing temperature to be regulated as desired."[103] Reviews uniformly praised the new hotel, albeit with some concern for the inequalities evident in this "Paris of the East." Movie star Douglas Fairbanks noted this contrast in 1930, writing, "With every heavy frost a dozen [migrants] are found dead in its streets and ten minutes ride from the Cathay Hotel, a sixteen story hostelry that has few equals in the world, hundreds of thousands of Chinese are living in squalor that beggars description."[104]

---

tivity not usually experienced by tourists is a visit to a food and produce market, led by a hotel chef. There is also a visit to a Chinese doctor who will explain the philosophy of Chinese herbal medicine."

101 Hongkong and Shanghai Hotels Ltd., 10.

102 For a popular history of Sassoon's age and the "life" of the Cathay Hotel, see Taras Grescoe, *Shanghai Grand: Forbidden Love and International Intrigue in a Doomed World* (New York: St. Martin's Press, 2016). See especially pages 88–99 for descriptions of the Cathay's architecture and other contemporary buildings at the time in Shanghai.

103 *Perth Daily News*, "An Eye Opener," November 15, 1929, 6.

104 Douglas Fairbanks, "Our Trip Around the World: Shanghai Days and Nights," *Cairns Post*, October 15, 1930, 11.

Fairbank's unease was shared, if not surpassed, by nationalist Chinese authorities and citizens. Flamboyant entertainment added to the "anything goes" ambience.[105]

While the Cathay Hotel hung on during the Second World War, it ultimately ran up against the ideology of the Communists as the 1940s ended.[106] *New York Times* columnist Henry Lieberman noted that hotel conditions deteriorated months before Mao's takeover. "Despite cheaper rates," he wrote, "the famous Cathay Hotel has become a gray and gloomy monolith. It has shifted its handful of guests to a sister hotel where servants still outnumber guests by at least forty to one."[107] Ultimately, taxes and labor policies drove the Sassoon's out of business and forced the Kadoorie's to abandon their palatial home in Shanghai as well. The Sassoon's retreated to the Caribbean, while the Kadoorie's fled the mainland, leaving only a small network of properties in Hong Kong. The "China question" would return in Hong Kong some seventy years later.

Decolonisation: American Overtures and Asian Hotel Networks (1945–1998)

The age of Asian decolonisation (1947–1998) ushered in larger planes and the arrival of Western hotel chains. Quite simply, widebody jets made hotels for the masses a necessity. The only question was who would own and design the hotels? Americans and Europeans or Asians themselves?

The United States Department of Commerce and prominent trade organisations, including the Pacific Area Travel Association, encouraged hotel development

---

105 Dancer Lilian Green had been "highly successful at the Cathay Hotel in Shanghai where her fast-stepping tap dances were immensely popular." *Philippines Tribune*, "Famous Comedy Star to Appear at Manila Hotel," October 15, 1936, 15. Some dance troupes split time between the chains of the competing Kadoorie and Sassoon concerns. The Di Gaetano Girls and their famous "Casanova Revue," "last arrived in Shanghai on October 19, and played a very successful engagement at the Cathay Hotel. Leaving Shanghai on January 8 . . . they arrived in Hongkong two days later, and opened at the Hongkong Hotel on January 11, where they played for one month to a very appreciate audience." See *Philippines Tribune*, "Six Dancing Madcaps Now Appearing at Manila Hotel Pavilion," February 27, 1936, 8.

106 Sassoon's masterpiece did not escape the Second World War unscathed. As *The West Australian* reported, "Prominent in between [Shanghai's skyscrapers on the Bund], are the tall Sassoon House, the home of the luxurious Cathay Hotel, which with the Palace across the road, has suffered in the current fighting." See "The Wonder of Shanghai," September 11, 1937, 7.

107 Henry R. Lieberman, "Austerity Dims Shanghai's Glitter As Depression Slows City's Pace," *New York Times*, August 23, 1949, 1.

to boost American, European, and to a lesser extent, Asian travel throughout the region. PATA collaborated with Harry G. Clement, the most prolific American tourism consultant in the 1960s and 1970s. Clement was a quasi-public figure who worked with the United Nations following World War II, and then started a successful career as a tourism consultant for the firms Checchi and Company and Zinder and Associates. He also opened his own firm. Under contract with the Department of Commerce, Clement worked with the PATA to complete a study of how seventeen Asian nations could better attract American tourists.

Wherever he traveled, Clement articulated the same strategy for attracting loans from banks or aid agencies to construct luxury hotels.[108] He boldly predicted that East Asia and the Pacific needed as many as 30,000 additional hotel rooms by 1968 in order to keep up with the growing number of visitors.[109] Clement believed that hotel development, like his reorganisation plans for national tourism commissions, should proceed in an orderly, top-down fashion. Clement spurned the idea that local and national governments rely on the strategy of grading hotels throughout the country to encourage upscale lodging. The grading system, he thought "offer[ed] weaker incentive to build and run truly first class hotels, and it would merely provide travel agents, carriers, and tourists with information that might often be out of date."[110] In contrast, if national governments demanded minimum standards for hotel quality and also offered incentives to investors building to first-class specifications, those countries would be in the best position to increase tourism revenue.

For example, after citing the need for six hundred additional hotel rooms in Singapore, Clement urged that the government "make concessions and . . . provide incentives so that potential investors will put their money in first-class new hotel development."[111] Clement was also very insistent on the type of business arrangements that should sustain first-class hotel development. While visiting Hong Kong he noted that many hotels were already under construction. "Reportedly, there has been no establishment of minimum standards for quality, amount of public space, size of rooms, or caliber of operations," he noted.[112] While he stressed that incentives be offered to encourage higher quality construction, Clement went one step further, arguing that hotel builders and the management team for the finished hotel should work together. These specifications clearly favored large-scale chain

---

108 Harry G. Clement, *The Future of Tourism in the Pacific and Far East* (Washington: U.S. Department of Commerce, 1961), 55–66.
109 Ibid., 57–59.
110 Ibid., 104.
111 Ibid., 116.
112 Ibid., 103.

hotels like Hilton International and InterContinental Hotels, who had a history of working together with the builders of the hotels that they managed—as well as with the airlines that brought passengers to Asia.

Like Hilton International executives, Clement believed that an important part of the planning process, prior to applying for private or public loans, was to identify a "suitable site" and raise "'seed' capital to show that the group is seriously interested" in hotel development. Once these conditions were met, the local investors could approach local governments for loans or tourism incentive programs. He identified those sources of international financing that had looked favorably upon hotel construction, including the International Development Association, the International Finance Corporation, the Export-Import Bank of Washington, Colonial Development Corporation (United Kingdom), or the *Kreditanstalt fuer Wiederaufbau* (Germany).[113] This formula played a critical role in the genesis and expansion of both InterContinental Hotels and Hilton International. Clements' explicit suggestions that international investors and their governments approach the development process in the same way as American hotels chains themselves underscore the symbiotic relationship between recently decolonised governments, international hotel chains, and international aid agencies—and in some cases airlines.

Given the sheer magnitude of Clement's study, as well as the emerging economic transformation of Asia in the 1960s, it is difficult to precisely measure the impact of his work on the growth of regional tourism. There is no question that Asian tourism grew exponentially. The advent of jet travel—to a region that could have been considered the penultimate frontier of global tourism development (Africa being the last), played an unquestionable role in its' rise. Nevertheless, there is external evidence of his success.

Basil Atkinson, the one-time head of the Australian National Travel Association (ANTA), as well as president of the International Union of Official Travel Organizations (IUOTO), suggested that the report was important "not only because of the 293 page report itself—part of President Kennedy's drive as the 1960s being his 'decade of development'—but because of the time spent by Harry Clement and the Checchi team in (our) offices advising verbally and enthusing staff about opportunities and the role of tourism in their national future."[114] In the case of Australia, Atkinson argues, Clement's work assisted the Australian National Travel Association (ANTA) in convincing government officials that they needed closer cooperation with ANTA if the country were to reap greater rewards from tourism. On a broader level, At-

---

113 Ibid., 62–65.

114 Atkinson to the author, e-mail, November 4, 2009.

kinson also notes, Clement's study contributed to the liberalisation of currency laws that did away with taxes on Australians spending money abroad or foreigners sending money abroad.[115] As further proof of the impact of the study in Australia, ANTA followed up on the report by contracting J. Stanton Robbins, one of the principal team members of Clement's Department of Commerce consultation team, to conduct an even more in-depth diagnostic study of Australian tourism.[116] Finally, Atkinson observed: "I'd say that all PATA member-countries used elements from the Checchi study to improve their tourism plant. One close associate of mine was Major Harry Stanley, the then Executive Director of the Hong Kong Travel Association [(initiated by Lawrence Kadoorie with the encouragement of the colonial government in 1953)], who was an avid follower of the recommendations in the report. They stirred him to seek developers to hasten that territory's image as a prime tourist destination."[117]

To wit, Harry Stanley's Hong Kong Travel Association made express mention of the study and its implications for Hong Kong's future tourism development. "Hotels are a critical feature of the travel industry throughout the world, and particularly in Hong Kong," the HKTA's 1959–1960 Annual Report noted:

> The report leaves no doubt that visitors are being turned away for lack of accommodation in peak months . . . This was already the case during the year under review, and visitors are increasing at a rate which, if maintained, will double their number every three to four years . . . the most urgent need is for first class hotels, particularly in Victoria, and the announcement of plans for the erection of two such hotels in very welcome. These however can hardly be ready until 1963 or 1964, and even with them it appears that, at that time, the supply of high-class accommodation will still fall short of demand.[118]

At the time of the Clement study, Hong Kong boasted 2,500 hotel rooms, yet the consultants estimated that only "750 were considered suitable for tourists."[119] By 1970, according to the World Tourism Association, the number of hotel rooms in Hong Kong shot up to 8,567.[120] At least one government honored Clement for his work. In December 1966, the Korean government feted Clement's "'contribution to the development of tourism in Korea.'"[121]

---

115 Atkinson to the author, e-mail, November 8, 2009.
116 Harris, Kerr, Forster & Company and Stanton Robbins & Company, Incorporated, *Australia's Travel and Tourist Industry* (New York, 1965).
117 Atkinson to the author, e-mail, November 8, 2009.
118 Hong Kong Tourism Association, *Report of the Board of Management of the Hong Kong Tourism Association*, 1959/60, 13–14.
119 Clement, *The Future of Tourism in the Pacific and Far East*, 103.
120 World Tourism Organization, *Statistical Report on the Period, 1967–1976* (Madrid: World Tourism Organization, 1977), 210.
121 Checchi and Company, *A Proposal to Assist Alaska to Develop its Tourism Potentials*, 35.

More generally, Kenneth T. Young, former American diplomat in Asia, pre-sciently argued that the United States had been justified in its strategic, military approach in East Asia, but that Asia's emergence from the economic and political precipice demanded a new American approach. "No one nation – Asian or alien – will be able to dictate Asia's destiny or the key question of Asia's future organiza-tion," he observed, "Asian primacy will, indeed, replace foreign predominance." As a result, Young continued: "We should reduce our increasingly obsolete and much too visible overpresence [sic] in Asia. Instead, we should develop a new kind of relevant and modest partnership suitable to the Asian framework of this new Asian era." While not mentioning tourism development specifically, the Clement study fit within Young's recommendations for the immediate future of U.S. Asian relations. Young noted: "Asia's needs . . . will also require us to deem-phasize military hardware and to increase the right kind of political, social, and economic support for nation-building and region-building." This included, among other things, "a larger foreign aid program and better technical assistance, in both the public and the private sectors."[122]

American luxury hotel chains benefitted from the emphasis on building first-class hotels for international tourists. A 1964 article in the *Cornell Hotel and Restaurant Administration Quarterly* captured some of the energy in the interna-tional hotel sectors of Japan, Singapore, the Philippines, and Hong Kong. As Japan prepared for the Olympic Games during that same year, "a great many hotels have been built . . . and 9 more with 2,825 rooms and 5,146 beds are under con-struction in the city of Tokyo alone – all to be completed in time for the Games." In Singapore, InterContinental opened a new hotel, the Singapura. Not to miss out on the action, "Hilton is planning to build adjacent to the new Singapura . . ." Meanwhile, in Hong Kong, Hilton International and InterContinental Hotels went toe-to-toe to purchase prime land for a mammoth Hong Kong hotel. Hilton inter-ests prevailed, yet the Hong Kong Land Investment Company tapped InterConti-nental Hotels to manage their Mandarin Hotel. Finally, in the Philippines, Hilton International announced a three hundred-room hotel for Manila.[123] By 1970, over forty luxury hotels managed by InterContinental Hotels, Hilton International, Western International, Forthouse Trust, or Sheraton ITT were either in operation, under construction, or in development.[124] Sociologist Robert E. Wood reported that by 1977, "76% of the tourism investments of the International Finance Corpo-

---

122 Kenneth T. Young, "Asia and America at the Crossroads," *The Annals of the American Acad-emy*, 53, 60.
123 "Treasure Trove," *Cornell Hotel and Restaurant Administration Quarterly*, 1964, number 4, 30–41.
124 "Travel in the 1970s," *Cornell Hotel and Hospitality Management*, vol. 10, no. 2 (1970), 13.

ration have been for hotels with management or significant ownership by one of the following eight multi-national companies: InterContinental Hotels, American Express, [British Overseas Airways Corporation], Western International, Lonrho Ltd., Holiday Inns, Americana Hotels, and Oberoi Hotels."[125] After developing a strong relationship between the emergence of luxury hotels in Southeast Asia, state-directed tourism, and the assistance of supra-national institutions, Wood observed, "The heart of the mass tourism industry is the hotel sector."[126]

Synergistically, the hotel chains, consultants like Harry Clement, Asian Pacific businessmen, and governments began constructing the "Pacific Rim" through hotel development, expanding on what Christina Klein has denominated the "global imaginary of integration."[127] Australian tourism pioneer Basil Atkinson noted:

> I don't know the effect the [Checchi] report had on encouraging the big hotel groups like InterContinental and Hilton on their Pacific building programs but assume it must have been considerable because of the emphasis given to opportunities and future needs . . . John Payne, a Pan-American vice-president and well placed for the InterContinental chain development [was a firm advocate] for the commissioning and then the implementing of major recommendations in the Checchi Report.[128]

Conrad Hilton's philosophy regarding how hotel development served the strategic aims of the United States during the Cold War mirrored Harry Clement's application of modernisation theory. Comparing the state of the world in the 1950s and 1960s to a "great residence," Hilton observed that, "Its owners were white men who ran it for their convenience, comfort, and profit."[129] In decolonising Asia and Africa, Hilton continued, "Backstairs were the servants' quarters occupied by the yellow men, black men, even a handful of red men." Independence suggested, however, that "the walls are down and the little people of the world are standing up and demanding to be counted."[130] Charging "[that] communist gangsters have come forward to whip up and lead these little peoples of the earth towards self-government," Hilton suggested that his hotels offered technical assistance and an increased flow of tourists to these vibrant nations. The attraction of tourists to Asian countries would "[bring] . . . needed currency, greater understanding, and

---

125 Robert E. Wood, "Tourism and Underdevelopment in Southeast Asia," *Journal of Contemporary Asia*, vol. 9, no. 3 (1979), 281.
126 Ibid., 282.
127 Christina Klein, *Cold War Orientalism: Asia in the Middlebrow Imagination, 1945–1961* (Berkeley: University of California Press, 2003), 23–24.
128 Basil Atkinson to the author, e-mail, November 5, 2009.
129 Hilton, 9.
130 Ibid.

international goodwill," to the decolonised East.[131] In reference to the value of technical assistance, Hilton enthused, "What I propose is bulldozers instead of tanks . . . I propose starting a stream of good men around the world, scientists, technicians, doctors, crop experts, yes even hotelmen . . . Sharing our wealth of knowledge with those nations who do not yet possess it."[132]

Culturally, Conrad Hilton's vision of tourism development shared uncanny similarities to Harry Clement's formula. In 1963, Hilton International's publicity arm produced a travelogue of Conrad Hilton's demanding schedule for hotel openings, including three hotels in the Asian Pacific (Tokyo, Hong Kong, and Honolulu). The thirty-minute film highlighted the *Clementesque* dichotomy of modern mass travel, which included a show of faux tradition within the context of the modern comforts of a Hilton hotel. The film's narrator casually observed, "The traditional celebration in a modern setting is a symbol of Hilton's move into the future while preserving the exciting spirit of the past." In Tokyo, the Hilton hotel sat adjacent to a Shinto temple. Decked out in a blue robe and sporting a cylindrical rice paddy hat, Conrad Hilton took in the culturally controlled events at the hotel's June 1963 opening. The Hilton film also attempted to mediate the balance between modern conveniences and traditional landscapes by appealing to Conrad's own sentiments. The narrator observed: "Conrad Hilton shares the enthusiasm of all travelers. They want to see the world. They want to visit exotic places and move among foreign people, but they want to know that after sightseeing they can come back to a shower and a meal they trust. And prices they know are not inflated for the inexperienced traveler. Hilton's policy is to remove their worries and let them discover the pleasures for themselves." If the film failed to distinguish between the modern tourist and the traditional people she might encounter, it emphatically reinforced the non-participatory nature of mass tourism when it noted, "For the tourist, it's a chance to *see* and to understand other people and other places."[133] Hilton hotels conformed to this balance between the traditional and modern.[134]

Despite these overtures, Hong Kong remained at the center of *Asian* hotel networks with *Asians* directing much of the sector's activity. And, its primary architects, including the Hongkong and Shanghai Hotels Ltd.'s Lawrence Kadoorie, had their own plans for tourism development. As mentioned in chapter 2, Kadoorie, much like Goh Keng Swee in Singapore, sensed the growth of Asian tourism as acutely as that from the United States and Europe. Kadoorie not only saw this in-

---

131 Ibid., 13.

132 Ibid., 17–18.

133 My emphasis.

134 Hilton International, "Hilton: Innkeeper Extraordinary: Grand Opening Events at Seven Hilton International Hotels," 1963, Hilton Hotel Archives, University of Houston, Texas.

creasingly acquisitive customer base as an audience for cultural tourism, but also real estate and hospitality. The line between hotels and second homes was very thin:

> I cannot leave this [the subject of Chinese tourism in Hong Kong] without mention of the motor road which, though at present in an almost impassable state, connections connects Canton with Hong Kong. When the surface of this highway is improved it will become a main trunk artery, connecting Hong Kong not only with Canton but with many other points inland. At such time it is not unreasonable to envisage the development of the coastline along the Castle Peak Road with villas and country homes for those living in Canton and ever further afield.[135]

As far as The Peninsula was concerned, the fast-approaching age of mass tourism, with its jumbo jets and charter busses, certainly would bring with it big, boxy American hotels. This, however, would not superannuate The Peninsula, which cultivated ties with the aviation sector to remain at the head of the pack.[136] To compete in the jet age, The HongKong and Shanghai Hotels Ltd. executed several tie-ups with local airline Cathay Pacific to market the city through its signature hotel and carrier.[137] The two also collaborated on in-flight food service which likely redounded to the prestige of The Peninsula as the caterer of choice to Hong Kong's airlines at Kai Tak.[138] The joint enterprise catered to twenty-two airlines calling at Kai Tak by the mid-1970s and brought new meaning to the appeal of a stopover, as a company newsletter attested:

> Many world travellers agree that meals on flights out of Hong Kong are superior to those out of major cities throughout the world. Airborne meals have long been the subject of travellers' complaints and ridicule. One rarely has any choice but to accept the starchy, sleep-inducing fare offered on most flights. It therefore comes as a pleasant surprise to be presented with an attractively arranged tray of fresh food on one's flight out of Hong Kong. First-class passengers enjoy meals comparable to those served in Gaddi's, The Peninsula's award-winning gourmet restaurant and the food in economy class, though not as extravagant is just as good.[139]

---

135 Lawrence Kadoorie to Sir Patrick Abercrombie, Memorandum, "Possibilities of Hong Kong as a Tourist Centre," November 28, 1947, Hong Kong Heritage Project.

136 *Tiger Standard*, "Asia Must Plan for New Wave of Tourism," November 1, 1971, "Hong Kong as Other's See It," Folder H.15 A-3, 1963–1975, E05/01, SEK-10-002, Hong Kong Heritage Project.

137 *The Peninsula Group Magazine*, "The Best Way to Travel in South East Asia," number 1 (April 1974), 11–14, Hong Kong Heritage Project.

138 *The Peninsula Group Magazine*, "Pie in the Sky . . . And Other Good Things from Swire Air Caterers," Number 10 (April 1977), 46–49, Hong Kong Heritage Project.

139 Ibid., 47.

Whether for the food or accommodations, The Peninsula was where discriminating travellers met, touting itself in the jet age as the "Crossroads of the World," as one advertisement declared. Cathay Pacific played a role in that success as well, as did the Peninsula in that of Cathay Pacific as the premium stopover hotel in the city. The Kadoorie's Hongkong and Shanghai Hotels Ltd. would later expand their holdings to Manila, Beijing, Singapore, Tokyo, Bangkok, New York, Chicago, Beverly Hills, London, Paris, and Istanbul.

Hong Kong also served as a launch and landing point for other Asian hotel entrepreneurs. Singaporean based tin magnate Madame Wong-Mah Jia Lan, for instance, conducted much of her work in Hong Kong. She and her husband typically stayed at Asian-friendly guest houses. Sensing an opportunity, Wong-Mah purchased land on Nathan Road in Kowloon and constructed an imposing twenty-eight story edifice. Christened the Fortuna Hotel, it opened in 1963 with the express purpose of offering a five-star experience at four-star prices. Catering to an Asian clientele, the Wong's and their investment partners (from Malaysia, the Philippines, and Hong Kong) introduced features that imitated hotels like the Peninsula, including in-bathroom telephones and heightened attention to guest services (i.e., a high ratio of luggage porters to customers and personalised interactions). Restaurants catered to the palates of highly discriminating Asian diners, with Taiwanese dishes featured on the menu.[140]

If Wong-Mah found inspiration for a successful hotel in Hong Kong, Lien Ying Chow hoped to recreate the opulence of the Hong Kong Mandarin in Singapore. While Lien had previous experience as an investor in hotels in Singapore (the Ocean Park and Cathay Hotels), as well as resorts in Fiji, he tried a more hands-on approach in the 1960s and 1970s with the management of the Singapore Mandarin (no relation to the Hong Kong hotel of the same name). The Singapore Mandarin represented Lien's bullish view of the city-state's tourism potential. At seven hundred rooms, with an option for additional rooms in an adjacent annex, Lien's hotel brought together the best of Western hotel management with the preferences of a largely Asian clientele.[141]

While Lien used his own experience visiting leading hotels to identify competitive practices, he entrusted the hotel's initial management to his oldest boy, Sonnie, who had trained at prestigious hospitality programs at Cornell University and in Lausanne, Switzerland, prior to taking a role at one of the Hilton International hotels in Hawaii. Under his father's direction, Sonnie recommended the

---

140 Wong-Mah Jia Lan, oral history, series: Women through the Years: Economic & Family Lives, Accession Number 003564, 24 reels, NAS.
141 Lien Ying Chow, *From Chinese Villager to Singapore Tycoon; My Life Story*, with Louis Kraar Entrepreneurs of Asia series (Singapore: Times Books International,1992), 95–100.

construction of an Asian tourist friendly shopping arcade flush with high-end jewelry stores, a harbinger of the luxe shopping culture that sprouted around the Mandarin in subsequent years—known to shoppers the world over as simply "Orchard Road." The innovative revolving restaurant atop the hotel also appealed to Asian guests as something of a "first" in Singapore. It was something they might associate with upper echelon offerings in Hong Kong.[142]

Robert Kuok, known to some as the Malaysian "sugar baron," best represented the intersection of overseas Chinese investment and know-how in the hospitality industry. He was the founder of the Shangri-La Hotel Corporation. Raised across the Straits from Singapore in Johor Bahru, Malaysia, Kuok credited his Fujian-province (China) heritage with paving a path for capital accumulation. While European administrators established a rules-based system, it was the Chinese that built it up, he asserted in his autobiography. "The majority of the Overseas Chinese are moral and ethical people who practice fair play and possess a sense of proportion," he observed. "I will concede that if they are totally penniless, they will do almost anything to get their first seed capital. But once they have some capital, they try very, very hard to rise above their past and advance their reputations as totally moral, ethical business[people]."[143]

Prime property close to Singapore's Orchard Road fell into Kuok's hands. With the help of Lien's Japanese architect, Yozo Shibata (given name first), Kouk's group opened the first Shangri-La Hotel in 1971. It was a modest twenty-two story tower of five hundred and twenty rooms. Shibata was in the early stages of his career, adapting modernist architecture to Asian tastes (most notably at Bangkok's Dusit International). His early structures featured modernist buildings configured in tasteful geometric designs, accented with terraced water features, and lush landscaping. He founded the Kanko Kikaku Sekkeisha in 1966 and became the most prolific hotel and resort planner in Asia. As his group emphasized on its fiftieth anniversary, KKS "thrive[d] in Asia." While this element of their philosophy has something of a public relations slant, it in noteworthy nevertheless:

> The twenty-first century is being call the 'Age of Asia.' Since KKS began business in Asia in 1966, it has targeted the whole of Asia as its area of operation. KKS has handled many projects starting in Indonesia, followed by Thailand, Singapore, Malaysia, the Philippines, Taiwan, Vietnam, Myanmar, and currently China. KKS also has plans to work in India and other Asian countries in the near future.[144]

---

142 Ibid.
143 Robert Kuok Hock Nien (hereafter identified as "Robert Kuok") with Andrew Tanzer, *A Memoir* (Singapore: Landmark Books, 2017), 279.
144 KKS Group, "Philosophy," https://www.kkstokyo.co.jp/en/philosophy/, accessed June 10, 2022.

Through 2020, the group listed hundreds of architectural and interior design projects throughout the region, achieving renown primarily, but not exclusively with its Asian clients, like Lien and Kuok. International clients included "Hilton Hotels, Marriott Hotels, Four Seasons Hotels, InterContinental Hotels, Fairmont Hotels, Swissotel, and Kempinski."[145]

Kuok took the advice of the architect, Shibata, and arranged a management agreement with Western International (later Westin Starwood) executive Edward Carlson, though not after some disagreement over the Vietnam War. Once Kouk looked past Carlson's constant redbaiting, the two settled into an arrangement that lasted from 1971 until 1983 at both the Singapore Shangri-La and Kuok's second hotel, the Kowloon Shangri-La. As he observed in his autobiography, "Hong Kong was a much bigger pond than Singapore or Malaysia. I began to see very clearly that the CEOs of the top American, Japanese and European corporations were visiting Hong Kong, if not once a year, then once every two or three years. The senior VP would go to Singapore and the VPs or departmental managers would visit Kuala Lumpur."[146]

Shangri-La met high standards for elegance and polish, but were not garish, a nod in part to Kuok's application of his management theory to his hotels. Asian hotels, as we have seen in all three cases of Wong, Lien, and Kuok, were spacious and anticipated increased Asian travel. "One has to take a blend of the crisp Swiss style of management and the American factory style of management, where they have hotels with 1,000 or even several thousand rooms," Kuok observed. "To my mind, the optimum scale is somewhere between 800 and 1000 keys."[147] In addition to a number of resorts in Malaysia and Fiji, Kuok's investment group built "big city hotels, including Shangri-La Kuala Lumpur . . . Shangri-La Bangkok, then the Island Shangri-La Hotel in Hong Kong, and many hotels in the major cities of China from the 1980s."[148] Modest elegance might have been the mantra of a Robert Kuok managed property. "Shangri-la hotels are dressed up smartly to qualify for a five-star hotel appellation, but they are not ornate or fancy. I don't build dream castles."[149] But location and scale were paramount:

> If you can build an 800-key hotel sitting on one of the crossroads of the world, you can achieve a high occupancy, which enables you to charge a rate commensurate with the value

---

145 KKS Group, "Service," https://www.kkstokyo.co.jp/en/service/, accessed June 10, 2022. An entire list of the Group's projects since its inception in 1965 can be retrieved from https://www.kksto kyo.co.jp/en-works/archive/?id=1960#a1960, accessed June 10, 2022.

146 Kuok, 292.

147 Ibid., 224.

148 Ibid. 223.

149 Ibid., 225.

that you are offering. It's like being a fisherman. Where are the currents flowing that bring schools of fish? You must park your hotel where those schools of fish tend to swim. You shouldn't be fishing in an area of the ocean where there is no current and therefore no fish around.[150]

Kuok looked askance at awarding contracts for management to Western hotel chains after his experience ended with Westin in 1983. Thereafter, he undertook the training of his own hospitality management cadre. While Kuok's slowly growing empire of hotels were managed according to Western standards, they were refined to Asian tastes. This was most conspicuous in his most lavish property: the Shangri-La in London's Shard. Kuok hired Hong Kong based Steve Leung to work on the interiors, which integrated:

> Rich materials and finishes in gold, sky-blue and bronze, [which are] embossed, layered and intensely decorated. Emblems are the traditional flying birds, trees and cherry blossom. Finishes are glossy with deep patinas and there is an abundance of ornate open-work screens and rich timbers . . . Lighting is chandeliers formed of tumbling golden droplets and branching clusters of the re-imagined traditional lantern shape. Furniture is squat, chunky, and bow-legged . . . The names of the key spaces within Shangri-La have an Eastern significance. The patisserie on the ground floor is called Lang, which is Chinese for 'pathway.' Ting, the name of the restaurant on the 35th floor, is a derivation of the Chinese word for living room.[151]

Designs flowed from Asian to cosmopolitan at the Shard, not simply in the materials, but also in approaches to unique spaces. The bar of the 52nd floor of the Shard was conceived by Andre Fu, who "fuse[d] his Asian background with ideas developed during his upbringing in Europe, especially London."[152] Ultimately, regional influences were important, as noted in official company promotional materials, which proclaimed, "We proudly bring our Asian heritage to the world and share its cultural richness in all our properties." Notions and niches reflected the "many diverse and colorful cultures . . . of each Asian city . . . through the designs of our buildings and its unique interiors."[153]

If Cold War politics played less of a role than commerce in justifying pan-Asian hotel development, decisions on how to accommodate relations with China influenced Kuok's enterprises. Kuok's involvement in building hotels and mixed-use developments on the mainland began almost as soon as Deng Xiaoping an-

---

150 Ibid., 224–225.

151 John Desmond Ltd., "The Shangri-La at the Shard," https://www.johndesmond.com/blog/design/shangri-la-shard/, accessed June 13, 2022.

152 Ibid.

153 Shangri-La Group, "Our Heritage," https://www.shangri-la.com/group/our-story/our-heritage, accessed June 13, 2022.

nounced economic reforms. For his first mainland project, he restored an aging property at Hangzhou's West Lake. One challenge involved replacing the entitlement mindset of both workers and clients in relation to the property. Thereafter, PRC officials paid for services just as any other customer. Further, new efficiencies were implemented to tame the costs of managing the hotel. Kuok eventually won a bid to build the Beijing World Congress Center, complete with two hotels. This long-term project, which ultimately paid handsome dividends as China's economy flourished, also required utmost patience in working with party leaders unaccustomed to the subtle arts of negotiation. Kuok judged his involvement purely on China's efforts to achieve economic development. "When I saw what Deng Xiaoping was doing," he observed, "I virtually worshipped the man."[154] "I have often told overseas friends that, throughout China's 5,000-year history, there has rarely been a period when the leadership has been as committed to providing for the people and nation-building as that since Deng came to power."[155]

In retrospect, while U.S. hotel chains indeed placed large hotels in major cities, the Asian hotel had evolved, distinguished by two features during the age of decolonisation: a larger size—indicative of the blurred line between permanent residence offered by penthouses and condominiums and the more temporary nature of hotels, as well as incomparable levels of service. As Eunice Seng observes in her seminal article on modernist hotels in Southeast Asia, "The adaptability of the architecture from housing the temporary domestic functions of the hotel to the relative permanence of the domesticity of the home continues into the present time."[156] Such was borne out in the perceptive newspaper reports. In 1986, a column in the *Canberra Times* entitled, "Hotel Design is Different in Singapore," noted that "hotel rooms are normally larger than the standard size of most other countries and have all the usual amenities." Enhanced services matched the expanded scale of rooms vis-à-vis the rooms of their Western counterparts, often referred to as "shoeboxes." "Room service is available around the clock in most establishments," the *Canberra Times* continued, "Extensive recreational facilities are offered in major hotels, including a health club with gymnasium, sauna, massage and sometimes tennis and squash courts as well as swimming pools. Laundry, babysitting, secretarial services, doctor on call, tour desk and car hire service, drug stores and money changing facilities are available within most hotels, while some boast shopping arcades with boutiques of international repute." Surprising to some, particularly those from the West, food in the hotel restaurants, earned high marks as well.[157]

---

154 Ibid., 304.
155 Ibid.
156 Seng, 1124.
157 *Canberra Times*, "Hotel Design is Different in Singapore," May 11, 1986, 8.

In a pithier summation, a travel writer for the *Australian Jewish News*, observed, "Those who think of a hotel in terms of a bed for the night probably have not been in Hong Kong. Staying there, cures most from the 'All I need is a bed and shower' mentality, and spoils all for lesser hostelries at larger prices." Topline institutions included "the fabled Peninsula, with its fleet of brown Rolls Royce's, its gloved lift attendants, French toilet soaps, and welcoming pot of jasmine tea," not to mention "The Mandarin, boxy and plain from the outside, glowing with warmth and service, flowers and candles, space and leisure inside . . . voted among the world's finest." Other refined touches included "staff [sending] a birthday card; wear[ing] spotless uniforms and white gloves; remember[ing] a name." Relegated to the second tier were "Sheraton and Hyatt, Holiday Inn and Regent . . . InterContinental and Hilton International." In sum, the correspondent noted, "In Hong Kong hotels, the Chinese . . . have got it right."[158] At the same time, American travel writer Michael Carlton noted that "Asia's luxury hotels continue to surpass those of any region, especially when it comes to service."[159]

Global hotel ratings validated such claims. In 1985, for example, *Executive Travel* magazine rated Hong Kong's Mandarin at "the best hotel of the year for the third year running."[160] U.S.-based *Institutional Investor* ranked Bangkok's Oriental Hotel "tops in 1985," followed by the Hong Kong Mandarin, The Hong Kong Regent, The Tokyo Okura, the Shangri-La, Singapore, and the Hong Kong Peninsula in 7th place.[161] Rankings changed little from year-to-year. Three years later, the British magazine *Business Traveller* ranked the Singapore Shangri-La as the top hotel "with its five [hectares] of manicured gardens spreading over some of the most expensive real estate in the world." It had been the best two years earlier, only to have been beaten out by the Bangkok Oriental and Hong Kong's Mandarin in 1987. Shangri-La, like the city's Changi Airport, left little to chance in the "Battle . . . for the title of Best Hotel in the World." Robert Kouk's management team financed "a $10 million refurbishment of its Garden Wing, and it's looking towards next year with a further $12 million being spent on the lobby, coffee shop and guest rooms in the main building."[162] A more local assessment of Singapore's offerings in 1988 echoed Shangri-La's plaudits, but warned that "the Marina Mandarin, threatens this ranking." It went on to list several hotel offerings that would be foreign to Westerners, including "The [Peninsula Group's] Marco Polo, The Orchard, The Mandarin, The Meridien, The Pavilion InterContinental

---

158 *Australian Jewish News*, "Variety, Class in Hong Kong Hotels," June 14, 1985.
159 *Canberra Times*, "Regent of Sydney Up with Leaders," March 11, 1985.
160 *Canberra Times*, "Mandarin Named World's Best Hotel," May 24, 1985, 4.
161 *Canberra Times*, "Bangkok's Oriental Hotel Keeps World's Best Ratings," October 10, 1985, 7.
162 *Canberra Times*, "Rewards Lurk in Unhappy Holiday," November 13, 1988, 22.

and the older, but still elegant, Goodwood Park." In contrast, the hotel reviewer found "the Sheraton's rooms too small for my liking and the Hilton somewhat 'old and tired.'"[163] In the end, Asian hoteliers, sometimes with Western assistance, and always with a knack for elegance and superlative service, held their own on the global front.

Diversion: From the Sands of Dubai to Marina Bay Sands (1998–2019)

Hotel networks established fortunes and made travel throughout the continent a pleasure, as it had been on the Imperial Airways hotel network. However, these chains often failed to create memorable architectural icons. Thus, at the turn of the twentieth-first century, as the cities of the Indo-Pacific vied for international air hub supremacy, some turned to starchitects to design buildings that rivalled any imaginable hotel elsewhere on earth. Such were the origin stories of places including the Burj al-Arab in Dubai.

By the 1970s, what had once been the focal point of refueling in the Persian Gulf at Sharjah had shifted, along with the evolution of jet technology, to a more formidable stopover in Dubai, famed for its duty-free shopping as much as for its reliability as a West Asian entrepot (discussed in chapter 2). Dubai also gained renown as an oil export outpost, outfitted with a few hotels located not too far north of the city at Chicago Beach.[164] In 1980, Sheik Hamdan bin Rashid Al-Maktoum cut the red ribbon on a new 630-room resort at the beach, which, in later days was remembered by former youngsters from Abu Dhabi as a vacation away from the Emirates capital.[165] Along with its traditional souks and duty-free shopping at the airport, Dubai looked for new angles to attract a more substantial tourism base as the prospects for storing oil off Chicago Beach ran dry. A few brave souls stayed at the resort during Saddam Hussein's invasion of Kuwait in 1990, but for the most part, sceptics questioned Dubai's pretensions to global sig-

---

163 Walter Reisender, "Travel Tips," *Australian Jewish News*, October 21, 1988, 50.
164 *Gulf News*, "40 Years of UAE: Maqta Bridge, Chicago Beach, and the Dubai World Trade Center," October 29, 2011, https://gulfnews.com/uae/government/40-years-of-uae-maqta-bridge-chicago-beach-and-the-dubai-world-trade-centre-1.920060, accessed July 2, 2022.
165 *Gulf News*, "Hamdan Opens Chicago Beach Hotel in Dubai," October 14, 2020, https://www.pressreader.com/uae/gulf-news/20201014/281745566860164, accessed July 2, 2022; Ali Al Saloom, "The Ali Story: A trip to Dubai's Chicago Beach Remembered," *The National News*, https://www.thenationalnews.com/lifestyle/wellbeing/the-ali-story-a-trip-to-dubai-s-chicago-beach-remembered-1.391107, accessed July 2, 2022.

nificance, charging that it wanted to be everything to everyone.[166] As one journalist observed: "Dubai wants to be everywhere else in the world at once. It would like its ports to be the Singapore of the Middle East and the air-cargo terminal the Seattle of the Middle East; it wants to become a Miami-style cruise centre for the Middle East and it wants its golf courses to make the emirate the Wentworth of the Middle East."[167]

The tide slowly turned for Dubai in 1994 as travel experts tabbed the city "the world's most desirable holiday destination."[168] City-state leaders then floated the next strategy with a plan to create an iconic destination hotel, whose opulence they effectively kept under wraps during its three-year construction period. The property, announced to include a wave-shaped resort with 600 rooms and a tower with 200 suites, was calculated by one British trade magazine to help Dubai "cast off [the] stopover tag."[169] For Sheik Mohammed, the opportunity, "symbolized his pride as an Arab." Journalist Jim Crane elaborated further on the cultural significance ascribed to the hotel. "It would cement his legacy as one of the great Muslim builders. The sheikh's would become the most significant Arab monument since the Alhambra, built in Spain during Muslim rule in the fourteenth century. He would call it the Tower of the Arabs."[170] Little was revealed about the paradigm-shifting nature of the project, but the size of the suites gave some clue as to its ambitions (rumored to be from between 1,500 to 6,000 square feet).[171] When it opened in 1998, the blandly named Chicago Beach Tower was rechristened the Burj al-Arab, a self-proclaimed seven-star resort that became the icon of the emirate at least until completion on the Burj Khalifa, the world's tallest skyscraper (with its own luxury hotel, the Dubai Armani) in 2011.[172] An additional destination-based hotel, the Atlantis, was erected at the heart of the residential Dubai Palm in 2008. While this massive project mimicked an already existing

---

166 *Andrew Quinn*, "Gulf Beaches Still Lure a Few Fearless Tourists," *Reuters News*, August 29, 1990, Factiva.

167 *Financial Times*, "FT Traveller, Dubai 2 – Ideas from Everywhere," July 3, 1992, Factiva.

168 Simon Reeve, "Gulf's Holiday Secret is Out," *The Sunday Times*, October 23, 1994, Factiva.

169 Miller Freeman, "New Hotels Cast Off Stopover Tag," *Travel Trade Gazette UK*, January 7, 1998, Factiva.

170 Crane, *City of Gold*, 113.

171 *Middle East Economic Digest*, "MEED Special Report on the UAE—Chicago Beach—Landmark Set to Make a Bigger Splash," May 13, 1996, Factiva.

172 See the Burj al-Arab's architect's presentation on the creation of the hotel and its evolution as an icon for the city, https://static1.squarespace.com/static/519f87b2e4b02061e74db6b6/t/523c2a8ee4b0f0cfeaff5207/1379674766598/Creating+the+Burj+Al+Arab.pdf, accessed July 2, 2022. The goal, according to Wright, was to design an iconic structure that could be drawn in five seconds.

property in the Bahamas, Dubai's investors calculated ways to potentially replicate the Burj al-Arab around the world in places like Las Vegas.[173]

Though Emirates Airlines did not exploit the link between the hotel and Emirates in its advertising, the unveiling of the airline's audacious fleet of Airbus A380s flying around the hotel completed the connection between city, hotel, and airline. Dubai-based *Gulf News* reported: "The sense of expectation was palpable as the giant Airbus A380-800 came into view beneath the clouds and headed towards the Burj al-Arab. When the superjumbo swooped low near the hotel, the crowds lining Jumeirah Beach gasped with amazement – and cameras clicked by the dozen. . . . The Emirates-liveried A380 made about a dozen passes near the seven-star Burj al-Arab."[174]

As at Dubai, the leaders of Singapore reckoned that the elitist appeal of Raffles, not to mention the limitations of its size, required a more substantial offering which projected the city as a garden state. About the same time Dubai launched plans for the Chicago Beach Resort and Tower, the Singaporean Urban Redevelopment Authority floated guidelines for an ambitious mixed-use complex to be located across the Marina from the existing Central Business District.[175] While it would be a project to be appreciated and patronised by Singaporeans, the thrust of the project was to attract businesspeople and conventions to enhance the tourist landscape. Like the multi-project ambitions of Dubai, Singapore launched not one, but two projects, "[deciding] in April 2005 to proceed with plans for projects at Marina Bay and Sentosa [Island]. The resort at Marina Bay would be geared toward business visitors, while its counterpart at Sentosa would provide family-oriented attractions."[176] Finally, as at Dubai, the multi-use project at Marina Bay, which evolved into a tri-tower, three thousand room hotel complete with casino, dual theatres, a world class shopping arcade, a science museum, and marina promenade, "had to be an iconic design with world-class content."[177] What with the untimely global recession, as well as its sheer scale, the investor footing the bill for the project, Sheldon Adelson emphatically noted, "I can't reproduce that

---

173 Shane McGinley, "Seeing Double: Taking the Burj Al-Arab Global," *Arabian Business*, December 22, 2019, https://www.arabianbusiness.com/industries/travel-hospitality/435909-seeing-double-taking-the-burj-al-arab-global, accessed July 2, 2022.

174 Wilson, *Emirates*, 326–327.

175 Singapore Redevelopment Authority, "The Marina Bay Story," https://www.ura.gov.sg/Corporate/Get-Involved/Shape-A-Distinctive-City/Explore-Our-City/Marina-Bay/The-Marina-Bay-Story, accessed July 2, 2022.

176 Cheong Koon Hean, "Marina Bay: A New Waterfront for the Garden City by the Bay," in Anne Thompson and Christa Mahar, eds., *Reaching for the Sky: The Marina Bay Sands Singapore* (ORO Editions, 2013), 10.

177 Ibid., 11–12.

building. It's become the most important reference point in the world. I couldn't build one in another city. It would be immoral. This building belongs to the people of Singapore. It's part of their skyline."[178] A brief cameo in the popular film, *Crazy Rich Asians*, further launched the project into the global imagination as the symbol of twenty-first century Singapore—right after the the show's protagonists gushed at Changi International Airport's incomparable ameitities.

As the twentieth century gave way to a new millennium, and with the handover of Hong Kong to the Chinese mainland government, the chief entrepots of Hong Kong and Shanghai positioned themselves to best deploy their resources to business and leisure travelers alike. As had been its custom, Hong Kong retained its role as a top shelf conservative hub for Asian luxury tourism development. The Kadoorie-led Peninsula chain maintained the elite status of its eponymous flagship, while also expanding to discriminating cities throughout Asia and Europe. Mandarin Oriental, created by the tie-up between the Hong Kong Mandarin and the Bangkok Oriental, launched a clutch of hotels in Asia, as well as Europe, the United States (whose property at New York's Columbus Circle was designed by Moshe Safdie of Marina Bay Sands fame), and Latin America. To put it simply, Hong Kong was the source of unparalleled service and a launching pad for Overseas Chinese investment.[179]

Conclusion

This chapter by no means offers an exhaustive account of Asian hotel development, having passed over important players including the most prolific female hotelier in the region, Thailand's Than Pu Ying Chanut Piyaoui, who tapped family finances to open the Princess Hotel in Bangkok, a precursor to the famed Dusit Thani Hotel in the same city in 1970.[180] What this chapter does offer, however, is a model of the relationship between the development of commercial aviation and hotel networks

---

178 Sheldon Anderson, interviewed by Martin C. Pedersen, "The Making of the Marina Bay Sands," in Thompson and Mahar, eds., 22.

179 For a discussion on the diaspora of Hong Kong based hotel chains see, respectively, A.P. Dow Jones, "Hong Kong Hoteliers Investing Elsewhere," *The Australian Financial Review*, January 16, 1990, Factiva; and, Gillian Rhys, "Cradle of Hospitality," *South China Morning Post*, March 13, 2022, accessed July 2, 2022 Factiva. Jonathan Kaufmann provides a compelling, entertaining, and insightful contribution to the legacies of the Kadoorie and Sassoon families in his book, *The Last Kings of Shanghai*.

180 While there are likely more in-depth studies in the native Thai, the following give some sense of the family-owned empire started by Ms. Piyaoui in the 1940s: *Arabian Business*, "Why Shouldn't I Be the One to Build It," https://www.arabianbusiness.com/industries/travel-hospitality

in the twenty and twenty-first century. During the colonial period, stretching here from the construction of the first grand hotel of the East, the Great Eastern in Kolkata (1840) through the end of World War II, the historical hotels maintained their preeminence as stopovers. During this period Armenian and Baghdadi Jewish entrepreneurs flourished not only because of their business acumen, but also as cultural influencers in the old-world elegance for which many of the colonial hotels in Southeast Asia and Hong Kong would be remembered.

During the age of decolonisation (1945–1998), American tourism consultants, exemplified by the work of Harry Clement, promoted construction of large-scale modernist hotels built along the lines of U.S. hotel chains with aviation affiliations. The arrival of large hotels also marked the end of the relevance of the old colonial hotels, which fed off railway and ship passengers of an earlier age. New areas of tourist interest, such as the shopping arcades on Singapore's Orchard Road, further transformed these institutions, such as the Raffles property, into symbols of national identity. Ultimately, luxury Asian hotel spaces were typically more generous in size than their Western counterparts and offered better service. As a result, Western efforts to penetrate the market were successful, but did not dominate the region as they did in other emerging markets, including Latin America and Africa.

Lastly, with the construction of hotels like the Burj Al-Arab in Dubai, city-state entrepots serviced by long-haul airlines linked up with starchitects to build hotels that minted their cities as destinations rather than stopovers. Dubai and Singapore, especially, set the pace with iconic properties enshrined in the consumer imagination through movies (for Dubai see *The Wedding Party 2: Destination Dubai*, produced from a Nigerian point of view and *Crazy Rich Asians* for Singapore from a Chinese American perspective). There was no comparable film for Hong Kong, which remained a center for the Asian hotel diaspora and a benchmark for luxury hotel service, but, with the persistence of political protests (which crippled the Hong Kong International Airport), imposition of the National Security Law, and COVID (which took its toll on Cathay Pacific Airlines), the former colony lost some of its allure.

/why-shouldn-t-i-be-one-build-it-197075, accessed July 2, 2022. *The Nation* (Thailand) provides "Glimpse Into the Life of Pioneer Family Behind Tourism Boom," February 26, 2017, Factiva.

Chapter 4
Australia's Integration with Asia in the Age of Aviation

> The string of the traditional [air] links with Europe, North America, and New Zealand have weakened and [account] for a falling share of the trips. In contrast, countries in the rest of the Asia-Pacific region have recorded growth in share each five-year period.
> – Kevin O'Connor and Kurt Fuelhart[1]

> Wherever you go, a stopover acts as a holiday within a holiday, enabling you to experience a new culture, relax on a beach or shop up a storm— plus you get an extra stamp in your passport.
> – Darren Wright, Flight Centre Travel Group[2]

Introduction

The Kangaroo Route shaped and reflected Australia's relationship to empire and Asia. Unlike the Arab alternative to global aviation integration in the Gulf States or the entrepot model pursued by Singapore, Qantas' Australian network was both Western and intra-Asian. As such, Qantas's joint-operation of the Kangaroo Route with British Overseas Airways Corporation, beginning with the Brisbane to London Imperial Airways route in 1934, demonstrated loyalty to the Commonwealth system.[3] Qantas co-founder Hudson Fysh promoted this relationship well into the

---

1 Kevin O'Connor and Kurt Fuelhart, "The Asia Pacific Region and Australian Aviation," in David Timothy Duval, ed., *Air Transport in the Asia Pacific* (Farnham, UK: Ashgate, 2014), 79.
2 Darren Wright, Flight Centre Travel Group, in John Killon, "Pulling Out All Stopovers," *The Canberra Times*, November 25, 2017, Factiva.
3 This "special relationship" is reiterated several times in planning documents composed in the immediate aftermath of World War II. Hudson Fysh wrote: "During my trip [to London] I made every possible effort to push the interests of Australian overseas Air Transport which we are firmly convinced is bound up with that of the Empire, and to promote the truth that English speaking collaboration is essential to our future, and that of the Empire, and the rest of the World." A note accompanying a schematic of airline hierarchy within the commonwealth ran, "It should be an essential basis of all planning that B.O.A.C. and Empire Associates carrying the Empire's mail [and passengers] should have a monopoly of such carriage (a) against the foreign carrier, and (b) against rival British carriers, thus avoiding subsidies otherwise necessary and ensuring that surplus profits from air mail surcharges go to the Governments concerned and for their use." Hudson Fysh, Qantas Empire Airways Limited, No.1, "Planning for the Air Future. Qantas Empire Airways Offer Their Services," MP183/16, Exhibit 9, National Archives of Australia, Melbourne.

https://doi.org/10.1515/9783111326641-005

1960s. Later, Qantas' developed an intra-Asian system with deep ties to Japan, which underscored long-standing commercial links between the two countries. Efforts to establish an alternative stopover to Singapore in Bangkok, Thailand, also exhibited Qantas' growing integration with Asia. These later developments hinged on the instrumental work of John Menadue, former Japanese Ambassador (1976–1980), head of the Department of Immigration and Ethnic Affairs (1980–1983), and CEO of Qantas (1986–1989). Menadue calculated that Qantas' efforts would return the airline to profitability, set an example of Asian integration for other Australian companies, and break the stigma of "White Australia."

Framework

Breaking down Qantas' Australian model further, the nation's antipodal position, the pull of the British metropole, the intermediate role of Asia, and general economic prosperity all contributed to an unusually keen interest in aviation technology, airports, and the stopover experience. While Australians would rather just "get there," technical constraints strengthened regional ties between Australia and Asia. These ties shifted from Singapore during the colonial period; to Hong Kong in the age of decolonisation; and then to Southeast Asia more generally in the twenty-first century. Because of these technological limitations and Australians' global travel aspirations, airports, hotels, and layovers took on added importance in their travel experiences.

The intra-regional Asian geographic shifts suggested in this chapter are modeled on the scholarship of Kevin O'Connor and Kurt Fuelhart, who observe that "the origin of air transportation links between Australia and the Asia Pacific region lies in economic and cultural ties derived from British colonial structures, especially with New Zealand, Malaya (as then constituted [which would include and even emphasize connections at Singapore]), and Hong Kong." Later, advances in aviation technology spurred Qantas and Cathay Pacific to offer regular service between Sydney and Hong Kong beginning in the late 1940s. More recently, and because of what O'Connor and Fuelhart describe as greater economic integration with Asia, "Singapore, Thai, Cathay, Malaysia, JAL, Garuda, Korean Air Lines, China Southern and China Eastern aircraft are seen at Australian airports, providing services to cities such as Beijing, Shanghai, Guangzhou, Ho Chi Minh City, Seoul, and [Bali] – all of which played little or no part during earlier periods."[4]

---

4 Ibid., 74.

The entrepot function of Asian airports further strengthened Australasian relations generally, and travel connections more specifically, from stopover points including Singapore, Bangkok, and Hong Kong. It was these stopovers that intensified regional connections. As O'Connor and Fuelhart noted, "contact with Asians on extended stopovers to Europe, along with more exposure to Asian culture at home, and finally, the lower cost of access to resorts and cities closer to Australia, saw rising numbers of flights to Thailand, Malaysia, Singapore, Hong Kong, and Indonesia in particular."[5] Additionally, the infrastructure boom further cemented cultural integration. "These capacity increases," O'Connor and Fuelhart contend, "have served to underscore the hub function of cities with strong links to Australia such as Singapore, Hong Kong, and Tokyo, and enable places like Seoul, Bangkok, Shanghai and Guangzhou to accommodate increasing traffic."[6]

Thus, the entrepot experience nudged Australians towards engagement with Asia. While the research on stopover efficacy towards return visits is limited, some scholars have noted its efficacy.[7] The historical evidence from Australian travelers suggests that intermediate stops indeed encouraged Australians to engage with their Asian neighbors. Ultimately, prosperity enabled these encounters between Australians, Asians, and their antipodal cousins in Europe. As Ian W. McClean has persuasively argued, Australia stands as one of the only former settler colonies to attain standards of living as high as Europe and sustain that prosperity throughout the twentieth century.[8] Thus, Australians have earned the discretionary income necessary to travel over long distances not only for business, but also for pleasure. The constraints of technology have inadvertently made Asian encounters part of that process.

Historiography

Commercial aviation and tourism have played a vital role in the nature and scope of Australian-Asian relations. In the historical literature, no study would begin

---

5 Ibid., 76.

6 Ibid., 77.

7 Chuanzhong Tang, "Exploring the potential of hub airports and airlines to convert stopover passengers into stayover visitors: Evidence from Singapore," PhD dissertation, Department of Tourism, Sport and Hotel Management Griffith Business Group, Griffith University, https://research-repository.griffith.edu.au/bitstream/handle/10072/366161/Tang_2015_02Thesis.pdf?sequence=1, accessed January 17, 2023.

8 See Ian W. McClean, *Why Australia Prospered: The Shifting Sources of Economic Growth* (Princeton, N.J.: Princeton University Press, 2012).

without reference to Geoffrey Blainey's 1965 study, *The Tyranny of Distance*. As much a reflection of the Australian state of mind as a perceptive observation, the millennial revision of his classic study included a new chapter bringing the relationship between Australia and the "near north" into the new century. The arbitrary designation of Sydney over Melbourne as international aviation gateway, Blainey observed, made the former "the city of the yen and the greenback at a time when those currencies were increasingly powerful."[9] For purposes of holidaymaking, Blainey surmised, it was proximate time-zones rather than geographic distance that made Australia attractive to East Asians tourists.[10] Aviation advances, as well as economic and political ties, further hastened Australia's integration with Asia. At the same time, connections with Britain, though reduced theoretically to a twenty-one-hour flight, declined. Blainey found it ironic, however, that "the main cities of Sydney and Melbourne, along with the federal capital city Canberra, are still far away from Japan and China, South Korea and even Indonesia."[11] Innovations in aviation technology made time more elastic than distance.

Then Prime Minister Julia Gillard's cabinet-level white paper, *Australia in the Asian Century*, also cited aviation-linked tourism to intensifying the Australian relationship with Asia. The paper noted that during the 2011–2012 year, seven of the top ten inbound travel markets for Australia were Asian nations. Speaking of specific countries, the report observed, "Strong growth from India and Indonesia, supported by a younger demographic and expansion in other promising Asian markets, such as Vietnam, is expected to complement the growth from China."[12] New flight connections provided the medium for such contacts. "Increasing air services between Asia and Australia, as demonstrated by the recent expansion plans of Air India, China Southern, China Eastern and Jetstar, present a strong opportunity to build tourism links with Asia," the study ran.[13] On the whole, the study concluded, in the not too far future, approximately 45% of inbound tourism in Australia would come from Asian nations.[14]

Michael Wesley, then head of the Lowy Institute, prefaced his analysis of Australia's "internationalisation-insularity paradox" with a hypothetical traveler starting out on a journey from a gleaming terminal to an airport beset by disfunc-

---

9 Geoffrey Blainey, *The Tyranny of Distance: How Distance Shaped Australia's History* (2001), 365.
10 Ibid., 366–367.
11 Ibid., 369.
12 Department of the Prime Minister and Cabinet, *Australia in the Asian Century*, White Paper (Canberra: Department of the Prime Minister and Cabinet, October 2012), https://apo.org.au/node/31647, accessed September 3, 2022, 127.
13 Ibid., 128.
14 Ibid., 127.

tion. Reflective of the inversion of roles of Asia and Australia in the public imagination, he then writes: "Last night you were in Hong Kong, a crowded peninsular city with no assets other than its harbour and its seven million people; a city that has been invaded, claimed and counter-claimed five times over the past two centuries . . . a city now of gleaming, ostentatious success, booming energy and disciplined precision." By contrast, "This morning you're in Sydney . . . an inversion of our world has happened without us noticing:

> Australians have traditionally thought of Asia as poor, backward and unstable. When, or if, they went to Asia, they were used to leaving clean, sunlit streets, the latest technology and infrastructure, ubiquitous safety and prosperity, for an adventure among shabby high-rising, roiling street markets and exotic scenes. That Asia still exists. But another Asia has emerged, an Asia that showcases the future in the same way that America used to; an Asia that builds infrastructure with an ease that appears beyond our capacities here in Australia; an Asia through whose streets flows wealth that is eye-popping to Australians who have grown up thinking they lived in the rich, lucky country.[15]

Wesley argues that while prosperity brought Asia closer to Australia through unusually high levels of travel to the north, the country remains transfixed on domestic issues at a time when historically conscious neighbors looked to retake their place at the apex of global power. "It is only by taking responsibility for their fate in a completely new world that Australians will finally realise how lucky they are," Wesley concludes.[16]

The following year, pioneering Asian-Australian relations historian David Walker and emerging scholar Agnieszka Sobocinska released an edited collection of essays entitled, *Australia's Asia: From Yellow Peril to Asian Century*. An extension of both of their work, the compilation highlighted significant connections between the two regions stretching back into the nineteenth century. Walker's editorship of the collection built on his pioneering monograph, *Anxious Nation: Australia and the Rise of Asia, 1950–1939*, which established the long-standing nature of ties between Australia and China, Australia and Japan, not to mention other countries in Asia.[17] While only one of the book's selections touched specifically on Australian tourism, in Indonesian Bali, the authors addressed themes of Australian ambivalence to Asia, particularly in the context of relations with Japan

---

15 Michael Wesley, *There Goes the Neighborhood: Australia and the Rise of Asia* (Sydney: New South, 2012), 2–3.
16 Ibid., 174.
17 David Walker, *Anxious Nation: Australia and the Rise of Asia, 1950–1939* (Brisbane: University of Queensland Press, 1999).

and China, as well as outright exclusion of Asia in regional histories and the national narrative. It was in this series of absences that Ruth Balint observed, "When we think of Australia's relationship to Asia at its most elemental, we must think of the sea. Australia's oldest links with Asia are maritime."[18] Similarly, commercial aviation established new contacts between individual nations and Australia.

Sobocinska's later work on "people to people contacts" between Australians and Asians, including tourists, also hinged on widely accessible aviation. In summing up the broad spectrum of Australians who have interacted with Asians during the twentieth and twenty first century, she writes:

> Perhaps the most important single change came in the way Australians conceived of the relative distance between themselves and Asia. As the prolific Frank Clune enthused, 'there is no doubt about the aeroplane making short work of long distances' and air travel was particularly important in bringing the 'completely different world from which we have hitherto been isolated . . . within easy reach.' This forced a reconsideration of Australia's geographical and political place in the world, and Clune's voice was part of a growing chorus proclaiming that 'thanks to air travel, the Far East has now become the Near North.'[19]

The defining elements of Australian-Asian relations during the last hundred years as portrayed by Sobocinska infer the primacy of long-haul contacts as the catalyst of those encounters. The tenuous nature of person-to-person contacts, however, suggested only limited interactions between Australians and many Asians. The geographic scope of those contacts, stretching from Bali to Kabul and Kathmandu, all implicate commercial aviation as an immutable force in Asian-Australian relations.

The Australian Airport Obsession

As the discussion in the previous chapter illustrated, the Australian press demonstrated a nearly endless appetite for aviation narratives. This was matched during the twentieth and early twenty-first century by intense interest in international airports, driven in part by the search for creature comforts on layovers along the Kangaroo Route, but also as a reflection of the inadequacies of their own country's airport offerings.

London garnered special attention in Australia's Pre-World War II travel imaginary. Although Australia declared itself a federation in 1901, London remained the center of the overseas Commonwealth. As the historian Geoffrey Blai-

---

18 Ruth Balint, "Epilogue: The Yellow Sea," in *Australia's Asia: From Yellow Peril to Asian Century* (Perth: UWA Publishing, 2012), 347.

19 Agnieszka Sobocinska, *Visiting the Neighbors: Australians in Asian* (Sydney: Newsouth, 2014), 214.

ney has written, "On the eve of World War I, Britain was close to the peak of its power. Australians bathed in the warmth of the British sun. In many ways the two nations were one. Between them the flow of migrants, commodities, and ideas was usually smooth . . . Most ships in the ports were British, though Germany and France ran their own passenger liners."[20] Narratives of travel to the metropole featured in the press both before and after the advent of aviation. An anonymous correspondent for the *Melbourne Age* provided an overview of royal engagements and parties in London in August of 1887, noting the presence of distinguished Australians hobnobbing with royalty. Not to be left out, the author commented on a soiree thrown by the Archbishop of Canterbury: "I took advantage of my presence at Lambeth Palace to make an exploration of one of the oldest and most interesting edifices which the metropolis of the British Empire can boast."[21] Decades on, the *Melbourne Australasian* announced the approach of an engagement for eminent citizens of the Southern Lands: "Australians in London will have a first opportunity of seeing the quarters of the forthcoming Empire Exhibition on Monday afternoon, when Sir Joseph and Lady Cook will give a big reception in honour of the Prime Minister . . . at the Australian pavilion. Special trains will convoy the guests to Wembley, and afternoon tea will tie served in the Stadium."[22] A decade later, in 1935, travellers bound for the King's Jubilee Celebration overwhelmed available capacity by cruise liner. "The rush of Australians for London in connection with the King's Jubilee Celebrations is certainly exceeding expectations," the *Brisbane Telegraph* reported, ". . . The *Marella* [a steamship] has 185 tourists from Australia, including 53 booked at Brisbane."[23] Months later, a Miss McGovern held forth to a packed Literary Club in Katoomba, recounting her attendance at the grand event in London. "So far as the Jubilee Celebrations went[,] Miss McGovern . . . briefly described this magnificent spectacle with its glitter and excitement and brilliance of colors dominating the day," the *Katoomba Daily* reported. "The multitudes that took part in the procession and the hundreds of thousands of spectators would be ever an unforgotten memory."[24] Ultimately, commercial aviation showered the benefits of modern travel throughout the empire. The *Macleay Chronicle* made the apt relationship between long-haul travel, the blandishments of fine hotels, and its point of origin in the metropole: "Now that travel by air has become so popular with Australians it is interesting to note

20 Geoffrey Blainey, *A Shorter History of Australia*, revised and updated (North Sydney: Vintage, 2009), 169–170.
21 *Melbourne Age*, "A Sketch of Modern London," August 30, 1887, 4.
22 *Melbourne Australasian*, "Australians Abroad," Saturday, March 1, 1924, 52.
23 *Brisbane Telegraph*, "Rush of Tourists for London, Saturday, March 9, 1935, 17.
24 *Katoomba Daily*, "Travel Sketch," Friday, November 22, 1935, 4.

from the following London news the comforts provided on the routes to that city: — Cases of lavish comfort are appearing here and there in the most desolate wastes on the routes from England to India, Australia, and South Africa."[25]

Australia's chief aviation apostle, Qantas' Hudson Fysh, spoke at length about the delights of London. Fysh personified Samuel Johnson's aphorism, "Why, Sir, you find no man [or woman], at all intellectual, who is willing to leave London. No, Sir, when a man is tired of London, he is tired of life; for there is in London all that life can afford." As a result, Fysh observed, the uninitiated traveler often found themselves overwhelmed by the variety of activities there. "For instance, take London theatres," he mentioned, "I counted in my morning paper 55 theatres and music halls, leaving out picture shows. To see even half these in a period of two months would be a tall order indeed. It is the same in sight seeing."[26] One of Fysh's friends visited the city at the same time as the Qantas founder. During the traveller's initiation to the city:

> One simply sees what he is most interested in and lets the rest go, and this is why London is of unfading interest – there is always something left. I was amused at a friend of mine whose holidays fell due while I was in London. Though an elderly man of great experience, what did he do with the three weeks? He moved into London from his home in the suburbs to get to know the City better. I mean *interesting* London, its life, its galleries and museums, the old buildings, monuments and inscriptions.[27]

Personalising the message for his home audience, Fysh then observed, "I think to Australians the word 'London' strikes more magic to the heart than the name of any other city in the world."[28] Fysh then added a plug for Qantas and Imperial Airways service to the city: "Perhaps the best way to get to London is to fly there; it is quickest, it is the new way, and as each night is spent on l and much is seen. Imagine spending a late afternoon and night in the following places: Cloncurry, Darwin, Surabaya, Singapore, Bangkok, Kolkata, Karachi, Basra, Tiberias, Athens, and Marseilles."[29]

Following World War II, pilgrimage-like journeys from Australia to London continued. Architect Adrian Ashton's three-month tour of Great Britain's town planning drew significant attention, not only for what innovations might be offered for Australian housing patterns, but also for its historical significance. "My

---

25 *Macleay Chronicle*, "Travel by Air," Wednesday, October 18, 1939, 8.
26 Hudson Fysh, *Air Journey to England*, pamphlet, April 1938, Hudson Fysh Papers, National Library of Australia, 1, emphasis added.
27 Ibid., 2.
28 Ibid.
29 Ibid.

first pilgrimage was, of course, to St. Paul's, and what a thrill coming down Ludgate Hill, and free of the ugly steel railway viaduct that impaired the view, to see this great edifice towering serenely against the sky as it has done for so many generations of Londoners." Not long after, he transited London's ages-old arteries to Westminster Abbey, where "one is surrounded by English history from 1066 to the marriage of Princess Elizabeth." Such scenes were didactic as well as instructive of current relationships between England and Australia. "What great scenes of sorrow and exultation these grey and crumbling walls have witnessed," Ashton reflected, "and again how fortunate was the British Nation and Commonwealth that this other great symbol of her greatness and tradition did not go down into a heap of rubble." But Australia's future turned on innovation as well as an imperial past. Aviation centers were as rich with possibilities as venerable places of worship. Referencing his arrival in London by air, Ashton noted, "Airliners from all of the world were at rest on the tarmac or accepting or discharging passengers at this huge airport, which is in the process of being still further enlarged, no doubt to cope with the Brabazon's of the future." Shocked by the pleasure-seeking throngs adjacent to the airport, he "was told that this was just the usual crowd who on Sundays came out to [the airport] for the pleasant bus ride and to see the great planes coming in and departing." Thinking of home, he then observed, "When Kingsford Smith Airport is finally developed in Sydney and is linked to the city by electric railway system, Sydney will compare very favourably with the other great cities of the world in this respect."[30]

Because of numerous stopovers on early long-haul flights, Australian journalists and aviation experts followed the latest developments in airport infrastructure. Enthusiasm for European and American airports, including but not limited to New York, Berlin, Amsterdam, and London, reflected the geographic imaginary of Australian travelers. Hopes ran to hyperbole as one columnist in the *Moree North West Champion* wrote of the progress of airports, "In thirty years the surfaces of these early flying fields have already developed into huge areas, artificially drained, with concrete runways to bear the weight of the huge aircraft now in use." Of the staff and passenger amenities, the same writer noted, "Already these have grown into an elaborate collection of heated hangars and repair shops of aircraft, control towers, and administrative building for the ever-growing ground staffs, booking halls and restaurants for the passengers."[31] Common were articles from government officials, returning from London, who held forth on the state of airports in Europe. In September of 1938, for example, the Minister for Customs, a

---

30 Adrian Ashton, "An Architect Abroad," *Sydney Construction*, October 12, 1949, 2.
31 *Moree North West Champion*, "The Airport: Its Evolution and Development," May 11, 1939, 4.

Mr. White, returning from Europe, "said that he had been impressed by the great advances he had seen in airport construction. The Tempelhof airport, Berlin, and Schiphol airport, Amsterdam, were social centres and busy traffic termini."[32]

Deep interest in London's airports followed suit. As early as 1924, the *Brisbane Telegraph* walked readers through a day at Croydon Airfield. "It is the place where the magic carpet of 'The Arabian Nights' has become twentieth-century fact. Paris in two or three hours, Rotterdam in an hour and three-quarters, Brussels in two and a half hours, Cologne in three and a half hours. Prague in ten hours – that is the way they will whisk you away at [Croydon] if you have a mind for such an adventure."[33] *The Telegraph* captured the flurry of activity on the tarmac, as "every half hour or so a machine comes or goes." Up to 200 passengers a day would transit the field. "Slowly but surely the world which travels in taking to air."[34] Five years later, the *Maitland Daily Mercury* vicariously transported readers to Croydon, noting the integral nature of a landside hotel, where "one gets into closer contact with the cosmopolitan atmosphere. It is amusing to sit on an evening in the bar lounge, that has a very Continental appearance, and listen to the chatter of French, German, and Dutch breaking through the prevalent English."[35] A year later, *The Melbourne Age* reprised the theme of the terminal as "miniature seaport." The columnist noted, "In the building there is a hotel, where air travellers who are waiting to continue their journeys by other places can rest, and also a restaurant."[36] All of these insights were not merely good reading, but also indirect criticism of the glacial pace of aviation improvements in Australia. Frank L. McIlraith, a journalist, excoriated Sydney's wholly inadequate facilities for international travellers arriving at Mascot. Contrasting what he experienced in Sydney with the ultra-modern conditions at the Persian Gulf airport and hotel at Basra, McIlraith surmised, "If we could repeat, even in a minor way, at Darwin what has been done at Basra or in Batavia, we would have done something very tangible towards popularizing Northern Australia."[37] Instead, McIlraith lamented, engineers well past their prime continued to recommend facilities inadequate in both their technical precision and customer comforts.[38]

---

32 *Adelaide Advertiser*, "Airport Organisation in Europe: Praise by Minister for Customs," September 8, 1938, 22.

33 *Brisbane Telegraph*, "At London, Planes, Not Trains," July 24, 1924, 16.

34 Ibid.

35 Horace Thorogood in *The Independent* reprinted in *Maitland Daily Mercury*, "Air Travel; Fascinating Sights Croydon Airport," May 30, 1935, 11.

36 *Melbourne Age*, "Around a Modern Airport," August 7, 1936, 1.

37 Frank L. McIlraith, "Airports Shame Australia; Dutch Ground Organisation Far Ahead," *Sydney Smith's Weekly*, December 31, 1938, 3.

38 Ibid.

News of a new "world leading" airport outside London at Heathrow elicited sustained interest from Australian journalists. "London airport is rising from gravel pits, mud flats, ditches, and ponds of Middlesex to become one of the great crossroads of international air travel," noted the *Rockhampton Morning Bulletin*.[39] Early coverage quantified the scale of construction, beyond anything yet seen in the world in aviation, including a runway of 15,000 feet, capable of handling new aircraft weight requirements well into the future. At its peak capacity, it could cater to one hundred planes an hour when completed.[40] While Australian airports struggled to accommodate planes weighing less than 100,000 pounds, Heathrow would easily welcome new Bristol Aircraft's new Brabazon aircraft, with a passenger carrying capacity of 120 travelers and weighing in at 360,000 pounds.[41] The terminal was situated in between parallel runways with passengers delivered landside underground to avoid interference with the surrounding runways.[42]

As with its predecessors at Croydon, Schiphol, and Tempelhof, the post-war Heathrow attracted visitors with no intention of flying. "Many people with a spare Sunday take their families to see this great new centre which is laid out in holiday fashion," the *Busselton South-Western News* reported. "In the huge public enclosure, which looks out on the taxiway in front of the airport's buildings and gives a magnificent view of the 'planes as they glide down, there are sideshows and sandpits, restaurants and bars."[43] This democratic vibe signaled a new age in travel, as tourists supplanted elites as denizens not only of the terminal, but in the planes as well. "Their first impressions [upon arriving in England] are not of the white cliffs of Dover nor steaming up the Solent towards Southampton, but of landing at this unpretentious gateway in the middle of London's suburbia."[44] Additional passenger services included, "a medical centre, permanently staffed by doctors, three restaurants, including one with a dance floor, which will serve three-course a la carte meals, and numerous snack bars . . . hairdressing saloons, a newsreel cinema, a post and cable office, big lounges and a children's playground and indoor nursery under the supervision of children's hostesses." Anticipating Terminal 3 in Singapore, the roof would feature, "gardens sheltered from the wind . . . from which

---

39 *Rockhampton Morning Bulletin*, "London's Airport to Equal Any," June 14, 1947, 7.

40 *Mackay Daily Mercury*, "London Airport to handle 100 Planes an Hour," May 13, 1947, 4.

41 Ibid.

42 See *Rockhampton Morning Bulletin*, "London's New," May 21, 1949, 10.

43 *Busselton South-Western News*, "London Airport," November 5, 1952, 8.

44 Ibid.

friends will be able to wave good-bye to passengers . . . laid out with rockeries and fountains."[45] Service to Australia from the new airport commenced from May 1946, well in advance of Heathrow's formal opening.[46] *The Cairns Post* subsequently covered Prime Minister J. B. Chifley's arrival at Heathrow in July of 1948, noting that "he looked a trifle tired after the four-day flight."[47]

As mentioned above, Australians took early and unusually keen interest in all thing's aviation, particularly on long-distance journeys. For instance, Australian newspapers made note of airport hotels very early on.[48] Other articles featured tips on how to cope on long-distance flights and fight jet lag.[49] Perhaps of greatest significance, journalists and public figures alike needled the inadequacies of Australian airports Just as domestic airlines closed the loop for flying around the continent in an unbroken "chain," *The Sydney Sunday Times* called the local airport "shoddy enough to cause trouble with a daylight landing, but night approach would be out of the question." Night flying domestically would require significant improvements. "Mascot [(the official name of the suburb where Kingsford Smith International Airport was located)], as the principal landing ground for Sydney," the anonymous columnist continued, "must be brought up to date in the matter of surface, and then lighted so that night landing is not only possible, but encouraged. This would enable a continuous service to be instituted between Brisbane, Sydney, and Melbourne. Efficient lighting arrangement on 'dromes in other States would enable the entire round Australia trip to be covered in a few days."[50]

Post-World War II optimism for a new age in jet-powered aviation opened new horizons for Australian airports. While Darwin played an outsize role as Australia's first aviation port-of-call during the age of flying boats, The British Commonwealth Air Transport Council designated Sydney's Kingsford-Smith Airport in suburban Mascot as Australia's international gateway. This decision relegated Darwin to stopover status on flights between Australia and London passing through India, while Sydney issued flights headed to London's Heathrow Airport

---

45 *Sydney Morning Herald*, "London to Have World's Most Up-to-Date Airport," December 11, 1953, 2.

46 *Melbourne Argus*, "Australia Service from New British Airport," May 28, 1946, 1.

47 *Cairns Post*, "Australian Prime Minister," July 9, 1948, 1.

48 See *Rockhampton Evening News*, "First Hotel for Fliers; Berlin Leads the Way," 3 May 1929, 5. The first airport hotel in the United States opened in October 1929, reported the *Sydney World's News*, "Airport Hotel Opens," October 23, 1929, 6.

49 Roseanne Robertson, "How jet lag can dim the excitement of far-away places," *Canberra Times*, June 30, 1985, 66.

50 *Sydney Sunday Times*, "What is Wrong with our Airports," June 1, 1930, 11.

with service through Singapore.[51] Brisbane also lost out on the international airport sweepstakes, only retaining its domestic services.[52] Melbourne boosters were not too far behind their Sydney counterparts in advocating for international airport status. As one local booster wrote, in an article entitled, "Melbourne – Pacific Air Hub":

> The facts demonstrate . . . that probably no financial, industrial, or other wartime expert who came to Australia from across the Pacific made Sydney his business headquarters . . . The Australian economy is in the process of transition from a war to a peace strategy. If that fact means anything, it means that more and more Melbourne will tend to become and remain the proper terminal point for communications, whether by air or by sea.[53]

Furthermore, convoluted travel between New Zealand's South Island and Melbourne added up to 900 miles to the journey with a stop in Sydney, argued W.D.G. Robertson, editor of *Aircraft Magazine*.[54] Whether or not Robertson's plea reached the ear of legislators is moot since the Cabinet designated Melbourne an international airport well in advance of the 1956 Olympics.[55] Seven years later, municipal leaders and airport executives worked with Australia's domestic airlines and Qantas to make the proposed Tullamarine International Airport as appealing as Sydney's Kingsford Smith.[56]

Pundits were ambivalent towards the state of aviation infrastructure on the continent as World War II ended. Prior to the war, Norman Ellison, aviation writer for the *Sydney Sun*, judged that "Sydney's Airport is Now Only a 'Has Been.'"[57] Other airports on the continent were better prepared for international air service and Sydney's runways were too short to accommodate the heft of new aircraft.[58] Coincident with the ending of the war, a correspondent for the *Sydney Morning*

---

51 *Hobart Mercury*, "Sydney an International Airport Soon," October 29, 1945, 7. Sydney also retained the title of primary gateway for Pacific flights as well. See *Burnie Advocate*, "Sydney to Remain International Airport," January 12, 1948, 6.

52 *Rockhampton Morning Bulletin*, "Sydney is now International Airport," April 10, 1946.

53 *Melbourne Herald*, "Melbourne—Pacific Air Hub," February 11, 1947, 4.

54 W.D.G. Robertson, "Airport Should Be International," *Melbourne Herald*, January 19, 1949, 4.

55 *Newcastle Morning Herald and Miners' Advocate*, "Melbourne Airport Now International," February 15, 1950, 4.

56 Domestic carriers T.A.A. and Ansett-ANA requested separate, though identical domestic terminals on either side of an international terminal that would house Qantas operations in Melbourne; see "Melbourne Tullamarine Airport Panel, First Meeting, 30th September 1963," in "Melbourne Airport Tullamarine: Special Executive Panel for Planning of Terminal Complex, Department of Civil Aviation," 235/1/4. National Archives of Australia, Melbourne.

57 Norman Ellison, "Sydney's Airport is Now Only a 'Has Been,'" *Sydney Sun*, October 26, 1944, 4.

58 Ibid.

Herald dusted off the familiar trope of comparing Sydney's new airport with those in the United States and Europe. "There is reason for thinking that it will reproduce the main features of the great Idlewild airport which is being built in New York City."[59] While both included impressive terminals, regional variations were important. "If Sydney is to have an ideal airport, there will also be within the field well laid out gardens, in which passengers may walk while waiting to transfer from one plane to another or in which friends may await the arrival of passengers."[60] Intermittent hours for irregular flights landing during the small hours of the morning would be replaced by service "day and night."[61] In the new age, "There will never be an hour in which this giant air terminal in the Pacific will be without hundreds of workers on duty."[62] One of the great advantages of the Sydney field was its proximity to the city.[63]

The following year, BOAC sent out feelers for an around the world service that would take advantage of existing Royal Air Force routes linking Montreal and Sydney. At the same time, Qantas investigated the possibility of offering daily service between Sydney and London. While neither of these possibilities materialised in the short-term, the journalist responsible for the article made a connection between aircraft technology, around the world service, and airports, which they found to be wanting in Sydney. In bold print, the article entitled "Mascot to be Hub of World Air Line" pointed out, "Airline operators, providing the finest planes and services in the world, are forced to operate round the world services from an antiquated airport." As a solution, the writer asserted, "The least the Government can do is to get down to the job and provide an airport to match the quality of British round-the-world air service."[64] Years earlier, a disgruntled journalist, writing for the *Sydney Morning Herald*, penned an article entitled, "A Backward Airport."[65] "Compared with aerodromes abroad," they opined, "Mascot is ugly and backward." Appealing to the grandeur of Schiphol, Heathrow, and Tempelhof, the article ran, "An airport should be something more than a transport terminal. It should be a centre of public interest, a point at which popular support for aviation, which means so much to Australia, might be generated."[66] Comparing Europe and Australia, he continued, "Continental aerodromes frequently are

---

59 *Sydney Morning Herald*, "Sydney Mau Have Replica of Big U.S.A. Airport," August 9, 1945, 2.
60 Ibid.
61 Ibid.
62 Ibid.
63 Ibid.
64 *Sydney Daily Telegraph*, "Mascot to be Hub of World Air Line," February 5, 1945, 10.
65 *Sydney Morning Herald*, "A Backward Airport," September 1, 1937, 14.
66 Ibid.

favourite weekend rallying places. There are tea-gardens and bands, new planes are inspected, flights organised, and recruits are won to aerial transport."[67] In contrast, the author queried, "What happens at Mascot? The place is dreary at the best times."[68] Discriminating travelers agreed: "Mr. T.H. Eslick, world traveller, this week said that the aerodrome is at least 20 years behind the times."[69]

The Minister for Civil Aviation, T.W. White, led the charge to reimagine the continent's airports, pledging that "Great new airports which would avoid the mistakes made in the latest airports overseas were being planned for Australia."[70] White undertook this very public campaign by visiting airports throughout Europe; junkets featured in Australian newspapers in the wake of Sydney's designation as primary gateway. While Australian technology held promise in the field of radio-assisted air control, he conceded "Our Airports are Out of Date," in an article featuring a discussion of the multi-purpose uses of Europe's leading airports.[71]

White dialed in on the mad dash of cabaret-goers at Amsterdam's Schiphol Airport to watch passengers disembark from flights arriving from far off Batavia, Indonesia, before going "back to their tables enjoying a cabaret turn while waiting for the arrival of the Moscow plane."[72] White continued, "This is no fantasy of the future. It was actually happening before the war at Schiphol Airport, Amsterdam, and during 1949, more than 900,000 paying visitors added to this airport's revenue by coming to enjoy good food and good music while watching the great airliners come and go." Turning to examples in the United States, he noted that great streams of revenue redounded to public coffers and private fortunes by "cafes, florists, chemists' shops, and other concessions [which exceed] all other income yearned by some airports."[73] Ultimately, he argued, "Overseas airports are becoming an integral part of the cities which they serve, and are no longer merely machines for handling aeroplanes, passengers and freight. There is room in Australia at all principal airports for committees of interested citizens to cooperate with the airport management in improving and facilities."[74] Taken with the possibilities for customer comfort, White proclaimed that:

> the passenger terminal itself is perhaps the most interesting piece of engineering, because it must provide amenities for thousands of people and enable them to reach their aircraft

---

67 Ibid.
68 Ibid.
69 *Sydney Daily Telegraph*, "Wanted – A Modern Airport," April 28, 1939, 6.
70 *Sydney Morning Herald*, "Plan for Jets: Great New Airports," April 17, 1950, 3.
71 T.W. White, "Our Airports are Out of Date," *Melbourne Herald*, February 17, 1951, 4.
72 Ibid.
73 Ibid.
74 Ibid.

safely, quickly, without having to walk too far. It can be, in addition, a combination of terminal station, club, market place and recreation centre. Perhaps the most difficult thing of all is to try and look into the future and build airports which will still serve their purpose, although aircraft types may change.[75]

While housing demands temporarily eclipsed improvement of the terminal at Mascot, White touted the advantages of "the chance to study the best post-war terminal buildings in Europe, and in North and South America."[76]

Formal designation of Sydney's Sir Charles Kingsford Smith Airport as international gateway invited a handful of innovative suggestions for airport improvements geared towards arriving passengers.[77] "Plans cater not only for the needs of domestic airlines, whose activities are monthly reaching new peaks," one journalist surmised, "but will also serve overseas operators who are making our main airports finite important international junctions."[78] Passenger amenities on the drawing board included an "administration block [which] will contain aircraft control, meteorological, radio and other offices of the Civil Aviation authorities, together with a restaurant, bedrooms, lounges and complete hotel facilities. Linked with this will be a section containing a post office, bank, shops, hairdresser's salon, doctor's surgery and a news theatrette."[79]

Kingsford Smith Airport subsequently was one of the busiest airports in the world. The airport's acting manager, H.E. Fraser proudly pronounced, "It exceeds each of London's two main airports, Heathrow and Northolt."[80] With the inauguration of flight service from Pacific islands, Sydney's airport was more than a terminal destination on European and Asian routes.[81] "People who are kept awake at night or otherwise irritated by the noise of aircraft passing overhead from Mascot or taking off from Rose Bay are apt to wonder: Now where is THAT one going? They must make the best of the fact that Sydney is now the hub of air tourism in the South Pacific," wrote journalist Jack Percival. More precisely, he presciently observed, "Every year . . . Sydney, is attracting thousands of visitors from

75 Ibid.
76 Ibid.
77 *Brisbane Telegraph*, "Cabinet Approves 5 million (pounds) Sydney Airport Project," August 7, 1945, 3.
78 *Bowen Independent*, "Airport Progress: Australian Facilities to Reach World Standard," May 23, 1947, 7.
79 Ibid.
80 *Newcastle Sun*, "Kingsford Smith Airport One of the World's Busiest," February 3, 1951, 3.
81 Activity at the Auckland airport in the early 1940s conferred entrepot status on Sydney at the same time. See *Brisbane Telegraph*, "Busy Airport," August 19, 1940, 3.

Europe, America, and the Far East. They stay in town for a few days and then fly on to the Barrier Reef, New Zealand, Tasmania or Pacific islands."[82]

Planners anticipated this development for some time. Theoretically, the first step towards establishing a hub—even on the South Pacific periphery—involved linking Sydney with existing global networks of commercial aviation. "[T]he size of the airports in Australia which are necessary to handle international air transport are not related fundamentally to the size of nearby centres of population but must be designed as integral parts of international airways," wrote K.N.E. Bradfield, Acting Chief Airport Engineer for the Department of Civil Aviation. "Their location, type and dimensions must be related to the requirements of international air transport and their design is influenced by their position in the world's airways and by other airports in those airways."[83] In making a specific survey of the Mascot site as Sydney's most promising location for an international airport, Bradfield observed "[by] virtue of its geographical position and its importance as a city, it is expected that Sydney will be [the] stopping place on Pan American Airways connecting Australia to India and Europe, North America, the Islands north of Australia and the Far East, the Solomons, New Hebrides and New Zealand; it would be a junction of a number of these airways as well as an airport on a route circling the world."[84] As a South Pacific hub, Bradfield believed that like its American counterparts—namely New York's La Guardia Airport—Sydney's airport at Mascot could generate significant revenue from the non-flying general public that came out to see the goings-on of the airport, as well as flights themselves. To best prepare for this, he argued:

> A newsreel theatre would be included and it may be expected that this would be popular with a traveller who has an hour or two to wait at the airport for a connection in part of his journey, as well as with the visiting public. Ample galleries will be provided from which the public may watch the traffic and the general activities of the airport. In general, the building will be made as attractive as possible to the non-travelling public and it is hoped to derive considerable revenue from them.[85]

Tram, bus, and railway connectivity to the airport facilitated passenger access to the city, as well as maximised public accession to the facility to enjoy its many amenities.[86]

---

82 Jack Percival, "Sydney, the Hub of Air Tourism," *Sydney Morning Herald*, June 23, 1951, 5.
83 K.N.E Bradfield, "Report on the Development of an International Airport at Sydney," 1945, Department of Civil Aviation, MP391, s.5, Folder 3, National Archives of Australia, Melbourne, 12.
84 Ibid.
85 Ibid., 27.
86 Ibid., 13, 20.

Sydney's role as a hub gathered steam throughout the twentieth century. Tasman Empire Airways Ltd. started flying boat service between Auckland and Sydney in 1938.[87] Land-based planes replaced the flying boats in 1954. A "Coral Route" eventually linked Australia to Tahiti through New Zealand.[88] Jet service brought the two nations closer in the next decade.[89] Around the same time, Qantas purchased Fiji Airways. Based in Nausori, the small air carrier linked "flights to Australia and the United States."[90] Over time, Fiji emerged as its own South Pacific hub. "[Suva] is an excellent starting place for those who contemplate an extensive cruise of the South Seas," *The Balonne Beacon* offered in the Fall of 1953. "Such well known groups as Samoa, and Tonga are within easy flying or shipping distance, while also there are regular air and shipping lines to the New Hebrides, New Caledonia, The Cook Islands, and Tahiti."[91]

The New Zealand government had established Nadi, located on the opposite side of the island from the capital Suva, as a South Pacific hub following World War II. The airport, originally built by the United States during the war, played a surrogate role in New Zealand's fledgling aviation service. At Nadi, Air New Zealand transferred passengers from Auckland's airport to larger planes. What served as a refueling stop in the age before widebody jets transitioned to a hub with the rise of Fiji Airways. The third leg of the hubbing concept was also present as "the largest concentration of hotels in Fiji has grown up in and around Nadi." In addition to welcoming tourists from Australia today, it provides trans-Pacific routes to Australasia, including flights between the United States, Australia

---

87 *Perth West Australian*, "New Zealand Air Service. Inauguration for Christmas," September 1, 1938, 19. Service began in 1940.

88 *Queensland Times*, "New Zealand Comes Closer," July 13, 1954, 6; *Australian Women's Weekly*, "Tourists Join the Jet-Set," January 26, 1966, 27.

89 An article in the *Avon Argus and Cunderdin-Meckering-Tammin Mail* newspaper, entitled, "Sunny Tahiti—A Tourist's Paradise," May 15, 1952, 1, noted, "Long isolated by lack of regular transport, Tahiti is again opening up a Mecca for tourists. A regular air service of Tasman Empire Airways Limited is now in operation over what is known as The Coral Route, extending from Auckland (New Zealand), through Suva (Fiji) and Aitutaki (Cook Islands) to Papeete, Tahiti."

90 *Canberra Times*, "Qantas Buys Fiji Airline," April 19, 1958, 6. By the following year, shares had been split equally between the Australian and New Zealand governments. See *Beverly Times*, "Fiji Airlines," November 26, 1959, 7. After ownership of the airline returned to the Fijian government, the newly rechristened Air Pacific established "a five year agreement under which they will share traffic on Qantas flights between Australia and Fiji." See *Canberra Times*, "Fiji," July 15, 1982, 10. Qantas dropped its stake in Air Pacific at the start of the next decade. See *Canberra Times*, "Qantas Cuts its Stake in Fiji's Airline," July 12, 1990, 4. Tourism and Australian tourists transformed the islands economy. See *Canberra Times*, "Where a Friendly Welcome Awaits," July 26, 1992, 12.

91 *St. George Balonne Beacon*, "Fiji: Hub of the South Pacific," October 22, 1953, 6.

and New Zealand, as well as serving as an entrepot between the South Pacific and Hong Kong, Singapore, Tokyo, Beijing, Hangzhou and Taipei.[92] Prior to the outbreak of the COVID pandemic, the government aimed to establish the Nadi facility as a "'leading green airport in the Asia Pacific region.'"[93] The government's greater aim was to "evolve from the hub of the South Pacific to the hub of the entire Pacific Rim."[94]

Qantas also made long-term plans for east-west hubbing across the Indian Ocean to Africa and across the South Pacific to South America. Qantas launched service west across the Indian Ocean between Sydney and Johannesburg, South Africa in September 1952.[95] Plans had been in the works for at least five years. In November 1948, Qantas conducted a survey flight between Sydney and South Africa. Leaving on November 14, the Lancastrian plane touched down at Perth, Cocos Island, Mauritius, and Johannesburg six days later.[96] Greeted by the recently elected Nationalist Prime Minster, Dr. D.F. Malan, the two parties expressed warm wishes for future links between the two continents. Qantas representatives suggested a variety of reasons for initiating service. First, businesses favored the flights. "South Africa being mainly a primary-producing country, must import manufactured goods, and a market exists favourable to Australia for the sale of farm implements, tools, clothing, electrical equipment and general merchandise," the report outlined. Furthermore, "family connections" justified "fast travel across the Indian Ocean."[97] Finally, such a connection would further assist Commonwealth trade. "The opening of this proposed service would provide a new route for business men from England wishing to visit South Africa and Australia. This would then be possible in one round trip, such as London/Johannesburg, Johannesburg/Sydney and Sydney/London, whereas at the present time it virtually necessitates two journeys."[98] The Qantas reconnaissance team recommended service every two weeks, increasing to weekly service once demand justified higher frequency. On

---

92 *Wikipedia*, "Nadi International Airport," accessed October 18, 2022.

93 *The Phnom Penh Post*, "Fiji Aims to be Pacific Hub," May 6, 2019, Factiva.

94 Ibid.

95 Qantas flew trail runs between Sydney and South Africa during the summer of 1952, followed by regular service in September. See *Melbourne Age*, "Flight to Johannesburg Completed," September 5, 1952, 1.

96 Qantas Empire Airways Limited, "Report on Lancastrian Survey Flight Australia to South Africa," December 7, 1948, MP288/1/0, Air Routes/16, National Archives of Australia, Melbourne.

97 Ibid., 6.

98 Ibid.

the Pacific route, LAN established services between Santiago, Chile, and Sydney two decades later.[99]

Some critics, nonetheless, were not satisfied. A columnist for the sector-specific *Sydney Construction* issued a blistering attack on Sydney's perceived lack of progress. His point of departure was a damning comparison with a new facility in San Francisco, California. Citing cleanliness and favorable sanitation conditions as Sydney's only advantages, the columnist compared it to the fifty-million-dollar U.S. facility, where there is "space for 24 huge airliners of the Super Constellation type that Qantas uses on its Trans-Pacific service." As for the terminal to this new facility: "There is no congestion as they have individual floors for access. Luggage is handled through chutes which guarantee, designers claim, that cases do not become mislaid. The interior of the building is divided into airline passenger stations, ramps lined with stores, elaborate restaurants and coffee shops, and administrative offices." Further developments hinted at a proto-aerotropolis, as "a mall still to be completed on the highway approach to the building will contain a shopping center."[100] More central to the function of this Pacific gateway were maintenance facilities, which would generate additional revenue.

The columnist's vituperations bear repeating at length. Kingsford Smith had slipped in the global ranks to 12th by 1954, well behind the three million per year which coursed through the new San Francisco facility and near seven million through New York's three airports (Idlewild, La Guardia, and Newark):

> Mascot's facilities are primitive in the extreme, and the amenities . . . of Australia's leading air terminal would shame an outback airstrip. Passengers disembarking from any of the seven international and seven interstate airlines cannot be blamed for thinking they have landed slapbang [sic] in a poverty-stricken slumland that is in the throes of prohibition, a food shortage, and an overwhelming dearth of plumbers and decent sanitation . . . Overseas tourists – and for that matter, interstate passengers – search in vain for civilised amenities, such as a substantial meal. All meals served are in the 'light' category—that is, light in quantity, but heavy in cost. The hungry passenger has the choice of the ubiquitous sausages, pies, cakes, and sandwiches with quite unappetising fillings, washed down with very weak coffee, or soft drinks. And the passengers have to wait on themselves. There is no service of any sort in the lounge . . . As far as shopping facilities are concerned, the only shop that caters for the Overseas Terminal opens an hour before each aircraft lands, and closes an hour after it has gone. Quite often it is out of articles incoming or departing passengers need.[101]

99 Frank Cranston, "Airline Ahead of Qantas in Flights to South America," *Canberra Times*, May 10, 1973, 11.

100 *Sydney Construction*, "San Francisco's Elaborate Airport; Puts Mascot Right into Shade; Why Australia Loses Tourists," November 3, 1954, 6.

101 Ibid.

The author then expanded the point of comparison beyond San Francisco, noting Sydney's inferiority to "the ultramodern New York International Airport at Idlewild, hastily begun six years ago."[102] In contrast to the dearth of options at Kingsford Smith, New York's newest airport boasted, "two luncheonettes open round the clock . . . a bank, hairdressing salon, drugstore, flower shop, shoe-shine shop, men's and women's clothing store, a commercial photographer, and a camera and souvenir shop."[103] Not one to let up on his lambasting of Sydney's facility, he then leveled criticism at evening conditions, with approach roads "smothered in Stygian darkness, reminiscent of wartime blackouts." If San Francisco and New York's Idlewild were not enough, the author then described the delights of Washington D.C.'s new National Airport. Ultimately, what mattered most to this writer (and to many conscientious Australians), particularly with the approach of the Melbourne Olympics, was to "remodel the nation's 'front gates'—our airports and our harbour facilities—modernise them to conform with the accepted overseas practices which tourists naturally expect."[104]

A decade on, political sniping in Canberra centered on the need to bring the airport, "up to the standard required in a city such as Sydney."[105] Debate dragged on into the 1970s, whilst the Defense and Aviation correspondent for the *Canberra Times* encouraged the Minister of Civil Aviation to take the opportunity to travel abroad (as had T.W. White a generation earlier) to see the emergence of what today might be called an aerotropolis. "They have in fact become cities in their own right," aviation journalist Frank Cranston noted, "producing massive incomes and supporting huge populations. They maintain their own industries involved in the movement of people and goods and the care and maintenance of aircraft, electronics equipment, communications and even retailing and warehousing."[106] Dallas/Fort Worth in the United States, in Cranston's view, set the standard and offered examples of how to ease urban strains traditionally associated with airport expansion.[107]

Aware of its status at the end of the Kangaroo Route, but also cognisant of its role as a burgeoning hub with traffic from Fiji, Papua New Guinea, and New Zealand, Qantas approached the government about purchasing an interest in the

---

102 Ibid.
103 Ibid.
104 Ibid.
105 *Canberra Times*, "Sydney Airport Below Par," September 23, 1964, 12.
106 Frank Cranston, "Careful Planning Needed for Major New Airport," *Canberra Times*, May 25, 1973, 2.
107 Ibid. Cranston continued to advocate for progressive approaches to Sydney's existing airport as an alternative to a second metropolitan airport or even a replacement elsewhere on the east coast of Australia. See "Extension of Sydney Airport Suggested," *Canberra Times*, December 14, 1973, 12.

Wentworth Hotel, a downtown institution since the mid 19th century. Rebuffed in its' initial attempt, chairman Hudson Fysh made a more convincing case in 1950, securing blocks of rooms for staff and passengers.[108] Eventually, Qantas' seat capacity outstripped the number of rooms available at the venerable Wentworth, prompting the airline to conceive of a hotel worthy of the "jet age," also to be called the Wentworth. The new all brick, horseshoe-shaped structure boasted over four hundred rooms and held the distinction of a "destination worthy hotel," akin to Raffles in Singapore or the later Burj-Al Arab in Dubai. Twenty-thousand curious Sydneysiders stormed the hotel on December 14, 1966, as part of an open house run amok. Royalty and movie stars subsequently replaced the hoi polloi during the hotel's ensuing decades of splendor.[109] Qantas followed up its investment in the Wentworth with a joint venture hotel project with British Overseas Airways Corporation in Fiji in 1969.[110]

Hong Kong as a London Stopover

As airports evolved, Australia's air networks to Asia shifted after World War II. Prior to the war, Hong Kong did not make sense as a London stopover.[111] As noted in the previous chapter, Hong Kong served primarily as a terminal destination for Australian businesspeople and government officials prior to 1945. They considered the city-state key for business contacts and attended events including the

---

108 *Sydney Sun*, "Qantas Airways wants to buy Wentworth Hotel for Tourists," September 11, 1950, 3. Also see *Sydney Daily Telegraph*, "Qantas to Control Big Hotel," September 15, 1950, 11.
109 The development and opening of the hotel made the press in a variety of venues. See *Canberra Times*, "Qantas Plans New 4m. Sydney Hotel," October 6, 1962, 1962; *Australian Jewish News*, "New Hotel," September 17, 1965, 2; *Australia Women's Weekly*, "Hotel Settles into Sydney Scene," February 15, 1967, 16.
110 See *Canberra Times*, "Qantas -BOAC Hotel Deal," September 18, 1969, 23. "The Joint Airline investment announced yesterday," the article noted, "is designed to improve and expand facilities at the Mocambo Hotel and Nadi Airport and Fijian, the island's leading resort hotel."
111 Imperial Airways developed its London to Hong Kong service long after the London to Australia route proved possible. It hinged on flight service between Penang, on the London to Singapore route, and entered the testing phase in 1936. See *Brisbane Sunday Mail*, "New Hong-Kong; New Aerodrome Link with Australia," January 5, 1936, 12. Service was approved for operation by April 1936 (see *West Australian*, "Australia to China; Perth to Hong Kong in 6 Days," April 7, 1936, 18). China refused to grant Imperial Airways flyover access, thus lengthening flight time between Singapore through Penang and then onto Hong Kong (See G.W. Woodhead, CBE, "Flying over China; Hong Kong to Myanmar; New British Concession," *Sydney Morning Herald*, February 15, 1939, 16).

Empire Fair.[112] In sum, air connections with Hong Kong were tenuous in the 1930s.[113] Following Imperial Airway's route between Australia to Singapore, passengers could transfer in Penang to "a new air route . . . over the Gulf of Siam to Saigon in French Indo China, then on to Tourane on the east coast of Annam where a nightstop is made. The following morning the air traveller passes over the Gulf of Tong King, above the Island of Hainan, to land at Hong Kong, 'that most prized of all British possessions in the Orient.'"[114] Despite this option, which made travel from Australia to China possible within a week, ocean-going traffic continued to predominate.[115] The most plausible reason for viewing it as an aviation entrepot would be onward travel to mainland China or Japan.[116]

Advances in technology promised easier Australian access to Hong Kong. Imperial Airways pledged shorter flight times throughout the Empire in 1938.[117] Ever the aviation innovator, KLM pushed for flights made by Douglas DC4 planes as early as 1939, which rendered boat planes obsolete. This facilitated night flying and reduced flight time between Australia and Europe to three days.[118] These advances accelerated connections between Hong Kong and Australia. While some Australians anticipated a return to Singapore enroute to England, both Qantas, and the privately owned Cathay Pacific Airways, inaugurated service between Sydney and Hong Kong.[119] Qantas conceded reciprocal rights to a host of new international airlines in the age of decolonisation, including Cathay Pacific, a development that tested the Australian flag carrier's competitiveness.[120]

Hong Kong's stability was by no means assured following World War II. Communist forces ousted Chiang Kai-Shek's Kuomintang and imposed a new economic and social model that accelerated outmigration, particularly from glitzy

---

112 On Australia's trade menu and favorable trade with Hong Kong see *Melbourne Age*, "Trade with Hong Kong; Australia's Favorable Balance," July 25, 1933, 8; *West Australian*, "Hong Kong Market; Australia's record Trading Year," December 20, 1947, 11.

113 *Launceston Examiner*, "Air Link Policy," February 27, 1936, 10.

114 *Rockhampton Morning Bulletin*, "Australia to China: New Airways Extension; A Week's Journey," April 1, 1936, 13.

115 *Brisbane Telegraph*, "To Bring Vessels from Hing Kong to Australia," August 31, 1937, 12.

116 *Melbourne Age*, "Australia-Singapore-Hong Kong Air Service," January 6, 1937, 7.

117 *West Australian*, "Faster Air Services; Imperial Airways Plans," March 19, 1938, 23.

118 *Cairns Post*, "Big Air Scheme by Royal Dutch Lines; All Night Flights," February 18, 1939, 12.

119 *Adelaide News*, "Air Travel, Stopover Plans," June 4, 1946. Qantas Empire Airways managing director, Hudson Fysh noted, "Singapore was quickly getting back to normal. Hotels were reopening, Raffles were running again, and it was expected that by the end of the year it would be the pre-war Singapore again . . . because Singapore was so 'close to Australia,' he expected many Australians would be wanting to go there in the near future . . . Singapore was well on the way to being a great world centre."

120 *Launcheston Examiner*, "Competitive Era Returning in Airline Operations," January 17, 1950, 2.

Shanghai. "What is the future of scenically magnificent, commercially strategic Hong Kong: the only remaining safe refuge of both Western and Chinese business along the entire China Coast?"[121] queried a staff correspondent from the *Sydney Morning Herald* on March 2, 1949. British and Australian businesspeople worried about the stability of the colony, particularly considering fears that Communist forces might overwhelm the island, Kowloon, and the New Territories. Added to this were jitters expressed by "all the world's major airlines," who were concerned about the operation, much less completion, of Kai Tak Airport.[122] In July, a couple of journalists flew up to Britain's Far Eastern jewel on the first Qantas flight to Hong Kong. Brinsley Sheridan, of the *Melbourne Argus,* interviewed six business leaders who also had flown to Hong Kong on Qantas' maiden voyage. They were ambivalent, but cautiously optimistic. Many of their comments presaged conditions on the island after 2019. "Hong Kong is doomed," one offered, "Already Europeans are leaving, and more will leave, not all at once, perhaps, but in time."[123] Another anticipated a Chinese takeover once the embattled nation found stability, noting, "The average Chinese wants Hong Kong back and wants to see the Reds there, but not yet, because China has no economic stability at present."[124] John Loi's column for the *Melbourne Her*ald echoed this uncertainty, even as the city sparkled as "a show window for the west." He emphatically augured, "Nobody in Hongkong believes the colony will be attacked frontally by the Chinese Reds, but few are prepared to bet that the place will still be British in, say, 20 years time."[125]

These political misgivings aside, Hong Kong reigned as the Asian metropolis par excellence in the mid twentieth century. It has an essence often explained by reference to other cities, sort of like Dubai today. A Northwest Orient Airlines newsletter opined: "Hong Kong itself is a peculiar blend of London and Shanghai – double-deck buses and trolleys, English-made automobiles, Victorian building facades – and Chinese traffic cops who have all the mannerisms of the British Bobby, rickshas, pulled by coolies with their quaint conical hats . . . these are constant reminders to the visitor that he is in the Far East."[126] Alternatively, Christopher Rand of *The New Yorker,*

---

121 *Sydney Morning Herald*, "Hong Kong, China Coast Haven, Now Worries About Future," March 2, 1949, 2.
122 Ibid.
123 Brinsley Sheridan, "What is Hong Kong's Future?" *Melbourne Argus*, July 1, 1949, 2.
124 Ibid.
125 John Lui, "Hong Kong's Fantastic Boom Unchecked by Red Menace," *Melbourne Herald*, July 6, 1949, 4.
126 Northwest Orient Airlines, ""International Items," "How Others See Hong Kong," No. 10/55 (11/6), Folder H.15 A-3, "Hong Kong as Other's See It" 63–75, E05/01, SEK-10-002, Hong Kong Heritage Project.

wrote, "For those who like the place at all, Hong Kong is a paradise now – one of those rare cities, like Paris or old Peking, that excite even the most wayworn. I know a dozen ordinarily reasonable Westerners who have returned here for a visit in the past month or two, and all have babbled deliriously from the moment of landing."[127] Its' appeal derived decidedly from its cacophony of contrasts, culture and commerce, as another journalist summarized, "Without question Hong Kong is one of the most picturesque ports of call in the Far East. East still meets West here amid scenes of striking contrast. Hong Kong has its shabby, teeming tenements and its comfortable western-style dwellings, its mansion and its squatters' shacks, its European shops and its Chinese noodle stands, its ocean liners and its Chinese finishing junks, its modern automobiles and its rickshas."[128]

Hong Kong ascended in the Australian tourist imagination with Cathay Pacific and Qantas service from Sydney and cemented the city's status as a preferred stop-over for onward service to London. Mrs. L. Wright gave a succinct description of the city's setting in a talk to the Mount Gambier's Rotary Club, noting that, "The city, dominated by two bank buildings – (1) The Hong Kong and Shanghai Bank and (2) the Bank of China – huddles down on the waterfront, a fascinating blend of East and West. At night it is full of the color of neon lights, the blaring of car-horns and the never-ending clatter of Mah Jong bricks."[129] The preferred stop-over activity was shopping. Australian Mrs. W.T. Fountain succumbed to the blandishments of Hongkonger customer service which offered tea or beer while shopping. She marveled that "stores were open every day, including Sundays, until 9:30pm, and most people left their shopping until night, when it was not so hot."[130] During a stopover on the way to Japan, Major Brian Fleming, a member of the medical staff in Kure, reported that members of his cabin crew purchased a "silklined woolen suit, a camel hair coat, two nylon shifts, and a pair of shoes, all made to measure and delivered back to the hotel by midnight."[131] L. Wright mentioned that stores sold, "brocades, carved wood, jade, ivory, pearls, as well as a host of European and American articles" in shops which "stay open very late."[132] Shopping trips were also conducted vicariously, one columnist in

---

127 Christopher Rand, "Letter from Hong Kong," *The New Yorker*, March 27, 1954, in "How Others See Hong Kong," H.15 A-3, "Hong Kong as Other's See It" 63–75, E05/01, SEK-10-002, Hong Kong Heritage Project.

128 Henry R. Lieberman, "Hong Kong Still Welcomes Tourists," April 24, 1955, *New York Times*, in "How Others See Hong Kong," Folder H.15 A-3, "Hong Kong as Other's See It" 63–75, E05/01, SEK-10-002, Hong Kong Heritage Project.

129 *Mount Gambier Border Watch*, "Impressions of Hong Kong," June 1, 1954, 8.

130 *Sydney Morning Herald*, "Hong Kong a 'Heaven' for Shopping," November 1, 1949, 7.

131 *Burnie Advocate*, "Wealthy, Crowded Hong Kong," July 29, 1954, 9.

132 *Mount Gambier Border Watch*.

the *Australia Women's Weekly* noted, as "the requests of friends and family will have you rushing around that duty free city with a shopping list instead of a guidebook in your hand." Not to be outdone, the writer confessed that "like most passengers on a recent Cathay Pacific flight to Hong Kong, I had a list (but compared with some of the others in our group of six, mine wasn't bad). It read, camera lens, silk shirts, silver chain."[133]

Even though Cathay Pacific touted itself as Hong Kong's "hometown airline," and indeed enjoyed primacy in Asian routes from the colonial entrepot, British Overseas Airways Company and British civil aviation authorities held strict control over the Kangaroo Route running through its East Asian city-state jewel. This did not prevent Qantas, however, from exerting pressure on both Cathay Pacific and the British government to protect its rights on a growing spectrum of Kangaroo Route options. With the arrival of the jet age, European airlines clamored to join the increasing crowded skyways towards Australia. Lufthansa, Alitalia, Air France, and KLM all indicated some interest in operating the route to the Australian government. Canberra simultaneously pushed back against additional flights while also petitioning the British civil aviation authority for access to the route to London through Hong Kong.

At times this turned contentious. Based on colonial arrangements, for example, the British authorized Air Ceylon (Now Sri Lankan Airlines) to operate a route to London from Colombo. This created added competition for Qantas should the route be extended to Sydney. But the biggest challenge came from fledgling British upstarts, like British United Airways, which, in the early 1960s, applied for routes through Hong Kong to Sydney using discounted seating.[134] This was too much for the Australian government. It threatened to blow apart the tripartite agreement, that "magnificent piece of practical Commonwealth cooperation" between Qantas, BOAC, and Air India, which evenly apportioned flights and profits on the Kangaroo Route, if London acceded to British United's request without reciprocal rights for Qantas from Sydney to London through Hong Kong.[135]

Regionally, Qantas reasoned that since it was a government run airline and BOAC the "chosen instrument" of the British government, it could throw its weight around with Cathay Pacific on routes between Hong Kong and Sydney.[136] Aside

---

133 *Australian Women's Weekly*, "Hong Kong," July 5, 1978, 39.
134 "Summary Notes of Matters Discussed During UK/Australian Talks in London, November 27–30, 1961," BT 245/552, "UK/ Australia Air Services Agreement, Proposed BOAC /CPA Route, Hong Kong/Sydney, BNA.
135 Ibid.
136 See R.H. Oakeley, July 21, 1960, Confidential Memo, document 48, BT 245/552, "UK/ Australia Air Services Agreement, Proposed BOAC/CPA Route, Hong Kong/Sydney," BNA.

from its allegedly belligerent negotiators, Qantas upped the ante with Cathay Pacific by requesting additional rights beyond the three weekly flights it already operated between Sydney and Hong Kong. The Hong Kong government stepped in on behalf of Cathay Pacific to prevent added in-roads into Cathay Pacific's dwindling traffic on the route in question. In response, Qantas expressed its' intention to operate Boeing jet craft on the route while Cathay Pacific operated smaller, non-jet planes on the route. As a result, the Hong Kong government and Cathay Pacific called in the British civil aviation authorities to negotiate a settlement.

Subsequently, British administrators ruled that BOAC would operate flights on the Sydney-Hong Kong route until Cathay Pacific was able to compete with Qantas. British authorities also refused Qantas's bid for a route from Hong Kong to London amid the Qantas-Cathay Pacific impasse in 1961, even though they had originally planned to grant approval to the request.[137] Decolonisation stripped BOAC of its leverage in opening new routes throughout the former British Empire. Hong Kong remained one of the vital airports where it retained control over access to London.[138] Four years later, however, British authorities accommodated the Australian petition. Qantas cobbled together point to point traffic between Hong Kong and London with many intermediate stops.[139] Conversely, Cathay Pacific had to wait until 1980 to secure a route between Hong Kong and London. Initially, British Civil Aviation authorities rejected the Cathay Pacific application for access to a Hong Kong-Gatwick route but relented in the face of an outcry from Hong Kong officials. Cathay Pacific received authorisation along with British Caledonian to begin operations between the metropole and Hong Kong during the summer of 1980.[140]

---

137 Note 32, Folder "UK/ Australia Air Services Agreement, Proposed BOAC/CPA Route, Hong Kong/ Sydney, BT 245/552, BNA.

138 Alison Munro, "Discussions with Mr. D.G. Anderson, Director General of Civil Aviation, Australia," November 23, 1961, "UK/ Australia Air Services Agreement, Proposed BOAC/CPA Route, Hong Kong/Sydney," BT 245/552, BNA.

139 Qantas operation of a Sydney through Hong Kong to London service was inferred from advertisements running in Australian newspapers as of 1964. See *Canberra Times*, Tuesday, March 9, 1965, page 4.

140 Cathay Pacific's entire proposal to British aviation authorities to run a route from Hong Kong to Sydney, as well as its appeal of a negative decision, is in the archives of the School of Oriental and African Studies at London University. See Cathay Pacific Airlines, "Hong Kong London Route Application," JSS Box 3130, SOAS. The entire exchange over whether Cathay Pacific would win on appeal (which it did not, but the decision was subsequently overturned" is found in "Hong Kong/ London Route Application and Appeal," Box JSS 3025, SOAS. This box establishes CPA's credentials as Hong Kong's private carrier of choice within Asia, as well as documenting competing applications by British Caledonian and Laker Airways for the coveted route.

With a route to London established in the 1980s, Cathay Pacific looked to limit Qantas' ability to profit from intra-Asian routes linking Singapore, Bangkok, and Hong Kong, claiming that it deprived Cathay Pacific of regional traffic it was entitled to under reciprocal agreements between the city-state and Australian civil aviation authorities. The row was a central point of contention during formal negotiations between the two members of the Commonwealth.[141] Qantas finally emerged triumphant years after it had implemented an "Asia first" policy. The victory also coincided with Australian leverage in the relationship given Hong Kong's sliding reputation as a regional hub, due largely as a function of constrained capacity at Kai Tak Airport.

While Hong Kong remained a viable stopover or destination, three events raised Singapore's profile amongst Australian travelers. These included the incorporation of Singapore Airlines, the rise of Singapore's Orchard Road as a global shopping destination, and the construction of large luxury hotels, such as the Shangri-La.[142] While Bangkok made a play to replace Singapore as the Southeast Asian foil to Far East Hong Kong, the lure of duty-free goods and the latest fashions held greater appeal than the patina of antiques to be found in the shadow of a reclining Buddha at river's edge in the Thai capital.[143]

Return to Southeast Asia

Hong Kong remained an important destination for Australian tourists well into the twenty-first century, but as O'Connor and Fuelhart reported, Australian's "contact with Asians on extended stopovers to Europe, along with more exposure to Asian culture at home and, finally, the lower cost of access to resorts and cities closer to Australia, saw rising numbers on flights to Thailand, Malaysia, Singapore ... and Indonesia in particular."[144] Many of these locations held long-standing interest for Australians, stretching back to the onset of mass tourism in the early twentieth century. One such destination was Bali, linked by both boat

---

141 See *Canberra Times*, "Qantas to Push to Alter Hong Kong Air Deal," September 30, 1988, 14. The row persisted into the 1990s. See *Canberra Times*, "Aviation Talks with Hong Kong Planned," April 25, 1995, 17.
142 The Australian Department of Transportation and Commerce closely monitored the breathtaking rise of Singapore Airlines among Asian Pacific long-haul carriers.
143 Diane Armstrong, "Mention Singapore or Bangkok," *Australian Jewish News*, March 23, 1990, 6. Frances Berthelsen provides a dramatic recounting of a stopover in Bangkok, in "Stopover ... Bangkok," *Australian Women's Weekly*, July 10, 1968, 45.
144 O'Connor and Fuelhart, 76.

and KLM's interisland flight service (a carrier which would be renamed Indonesia Garuda following decolonisation). With a downturn in the sugar market in the late 1920s and early 1930s, a new crop—tourists— presented themselves in the form of motoring cadres of Australians, exploring what had been presented to them in their local press as an otherworldly and unchanging Hindu island people.[145] Three-day motor tours and cruise ship incursions into Balinese communities constituted a South Seas version of quasi-voyeurism practiced by tourists among indigenous peoples in Guatemala.

While journalists rarely missed the opportunity to highlight the "exoticism" of the Balinese, many tourists, Australian women especially, admired the artisanal skills of the island's women. Balinese men cultivated rice and managed irrigation systems. Tourists were not unaware that the "timelessness" of the island had been facilitated by the Dutch government, which provided the outlay for "fine roads and bridges" for visitors to "motor round the most populated and interesting parts with comfort in three days." What was unchanging was not the lives of the Balinese, but the tourists' demand for dances and ceremonies at temples. "They live mainly for their temples, cremations, and rice fields," was a tourists' refrain that changed little during the twentieth century.[146] Of equal interest to tourists over time were the communal ways of living, where "the spirit of cooperation amongst the agriculturalists" prevailed.[147] The thoroughgoing nature of religious observance provided a marked contrast to the generally segmented nature of life in the West.[148] Of greatest interest, perhaps, were the religious ceremonies, where "Warriors of the different tribes came from long distances – up to six days' travel – and gave savage dances to barbaric chanting."[149] Over time, the rise of mass tourism strained individual encounters, but did not stem the growing tide of Australians to Bali's beaches, a development facilitated by the Indonesian government and private luxury hoteliers alike.[150] By the late 1970s, Sinclaire Solomon of the *Port Moseby Papua New Guinea Post-Courier* declared the island to be the "Mecca of the World's Tourists Travel" where "every second person in tourist season is a foreigner in [Bali]."[151]

---

145 J.H., "Bali and its People Ready for Tourists," *Mackay Daily Mercury*, February 21, 1949, 6.
146 *Brisbane Telegraph*, "Island of Bali; Home of Beautiful Women; Tourists' Impressions," January 29, 1929, 15.
147 Ibid.
148 J.H., "Bali and its People Ready for Tourists."
149 *West Australian*, "Tourists at Bali; Visit to Famous Temple," April 16, 1934, 8.
150 In the post-World-War II era, Western tourists complained of the paper trail at Indonesian customs. See *Perth Daily News*, "Too Many Bali Forms Keep Tourists Away," March 1, 1951, 9.
151 Sinclaire Solomon, "Mecca of the World's Tourists Travel," *Port Moresby, Papua New Guinea Post-Courier*, February 2, 1979, 24.

Australia weathered the OPEC energy embargo and global inflation trends easier than other countries, including the United States and western European nations, due in part to deeper integration with East Asia. Moving past contentious issues including Japanese wartime attacks on Australia, Japan and the southern continent signed a trade treaty in 1957. By 1968 Japanese trade shares with Australia surpassed those with England. As Japan economy sizzled, tourist flows between the two countries increased.

Australian encounters with Japanese tourists transpired mainly in eastern Australia, within a complex relationship that goes some lengths towards explaining Australian preferences for stopovers in English-language friendly Hong Kong and Singapore. In 1991, as Japanese tourists made inroads in Australia (eventually surpassing New Zealand as the primary market some two years later), thoughtful journalists weighed in on the complex relationship, as well as offered suggestions on how to improve ties with Asia more generally. *The Canberra Times* published a feature-length interview with Bill Coaldrake, a research fellow in Australia National University's Department of East Asian History. Coaldrake pinned the problem on Australian's lack of language expertise, as well as the need to increase educational awareness of Asia generally, and Japan specifically. He urged restraint, particularly in the media, which often sensationalised Japanese investments in tourism as "threats." The learning curve went both ways. "The Japanese for their part are saying that they need to understand Australia better than just through tourism and kangaroos," Coaldrake observed.

Three years later, *The Canberra Times'* Beryl Cook interviewed Jocelyn Chey, Australian Consul-General in Hong Kong, about Australia's image throughout East Asia and steps that could be taken to remedy the label of "Ugly Australians."[152] Cook argued that Australian reticence to engage with its northern neighbors had less to do with racism than with a failure to engage as "both multicultural and distinctly Australian."[153] From her own experience growing up in Australia, Chey observed that seeing "'pictures of Borobudur'" in travel magazines "'really started me off with a realisation that Asian civilisation as a whole was a whole field of knowledge which wasn't included in what they served you up at school and a growing impatience with that insistence that you learned about British kings and

---

152 In an article published in the *Canberra Times* entitled, "Unpopular Tourists," the authors noted briefly, "The rising flood of Australian tourists through South-East Asia is not a pretty sight to the Asians. In fact, the ugly Australian tourist ranks as the second-least-popular visitor to Asia, slightly above the Japanese. *The National Times* exposes this threat to Australia's image in Asia." (Saturday, September 29, 1973, page 3).
153 Beryl Cook, "Asia Needs a True Australian Image," *Canberra Times*, August 1, 1994, 9.

queens. It was a determination to try and enlarge on those lessons.'"[154] While re-newed Asian immigration after 1973 rendered the continent a "truly multicultural society," perceptions of the "Great White Wall" continued to color perceptions be-tween Australians and Asians. Like Coaldrake, Chey encouraged moderation on the part of the media as well as updated historical and cultural curriculum in the schools. Those Australians and Australian businesses that achieved success, she noted, "[made] a name for themselves [with] a cultural approach."[155]

Qantas' Asian Integration

When, in 1992, Tokyo's stock market (the Nikkei) plummeted, the topic of rela-tions between the two countries filled the opinion pages of Australian newspa-pers. "Japan is our major trading partner," a columnist for *The Canberra Times* observed. "If the Nikkei's plunge is a precursor of a serious recession in that coun-try, the consequences for Australia are bad enough. Our exports of wool, iron ore and coking coal will be affected and, given the negative balance of payments that has haunted Australia in recent years, we can ill afford yet another blow to our trade."[156] Looking beyond the bi-lateral relationship, the author noted that trade with other Asian nations, including "Thailand, Indonesia, Singapore, the Philip-pines, among others," would also suffer as a result of contagion spreading from the Japanese downturn.[157] While it was believed that steady Japanese tourism would continue in spite of economic headwinds, additional investment in hotels along the Gold Coast and elsewhere, would likely dry up. Some saw this last devel-opment as proof that Australians had done little to deepen tourism ties to Japan, leaving it to outside investors.[158]

Nevertheless, the arrival of large numbers of Japanese tourists defined the late 20th century in Australia. Gold Coast airports expanded to receive growing numbers of tourists.[159] The Queensland government prepared for the deluge of tourists by printing copies of its tourist literature in Japanese in 1973.[160] Cultural

---

154 Ibid.
155 Ibid.
156 *Canberra Times*, "When Japan Sneezes," April 19, 1992, 7.
157 Ibid.
158 Ibid.
159 Ian Mathews, "Relying on Japanese," *Canberra Times*, October 18, 1972, 3.
160 *Noosa News*, "Japan" Brochure," May 10, 1973, 10.

exchanges between the two nations accompanied the onslaught of mass tourism.[161] Qantas steadily increased the number of non-stop flights throughout the 1980s to meet the demand, some of which was calculated "to make it cheaper for Japanese to fly to Cairns rather than to Hawaii."[162] Japan Airlines also increased its flights to Australia, even as the economy in Japan lagged.[163] Efforts in Australia and Japan bore fruit. Emma Conti, editor of the Subiaco, West Australian based *Japanese Perth Times*, predicted that tourism would continue to grow, having seen 528,500 Japanese tourists visit the island in 1991, fully 22.3% of the total number of tourists visiting the country.[164] The number increased to 629,900 tourists the following year, with Japanese visitors eclipsing New Zealanders for the first time.[165]

Qantas, under the direction of John Menadue (1986–1989), made its first concerted Asia strategy. Menadue's career as a public servant put him in a position to advocate for these changes. As a college student, Menadue boarded with three Malaysian brothers while attending school in Adelaide. During the Gough Whitlam administration (1972–1975), Menadue served as head of the Department of Prime Minister and Cabinet, an experience which included working on the repeal of the Immigration Restriction Act of 1901. Menadue developed a personal interest in Japan because of business relationships forged while working in the private sector for Rupert Murdoch. Family trips to Japan, including boarding at traditional *ryokans*, stirred a deep interest in the history and culture of Australia's main trading partner. Menadue subsequently served as Ambassador to Japan for five years between 1976 and 1980. His engagement with Asia did not stop there. He served as Head of the Immigration and Ethnic Needs Department between 1980 and 1983, a period marked by an influx of thousands of Southeast Asian refugees. "So, the experience with Qantas was all part of my experiences going back to my experiences in Japan, which Qantas hadn't accepted in the years when Japan was the economic wonder of the world," he told me. "There was a book written, *Japan as #1*, I believe it was. It didn't work out like that, later on, but Japan was then a boom country of the world and here was a great business opportunity and Australia and Qantas was not taking advantage of the opportunity."[166]

---

161 *Canberra Times*, "180 Japanese Tourists in Canberra," December 30, 1974, 6.

162 John Jesser, "Qantas Increases Flights to Cater for Japanese Tourists," *Canberra Times*, October 8, 1986, 8.

163 *Canberra Times*, "Japan Airlines Expects in Tourists to Australia," March 27, 1992, 11.

164 Emma Conti, "The Effect of Japan's Economy on Australia," *Japanese Perth Times*, June 1, 1992, 1.

165 *Canberra Times*, "Japan Sets New Mark as Source of Visitors," March 18, 1993, 9.

166 John Menadue, interview with Evan Ward, August 11, 2022, in author's possession.

Eager for a new challenge, Menadue accepted appointment as head of Qantas in 1986. In an interview with the author, Menadue observed that his previous experience in Japan emboldened him to spearhead an "Asia first" strategy. This consisted largely in expanding flight routes and frequencies between Australian gateways (including Sydney, Melbourne, Brisbane, Cairns, Adelaide, and Perth) and Japanese cities (two airports in Tokyo, and one each in Osaka, Nagoya, Fukuoka, and Sapporo). He also hoped to launch a new intra-Asian hub in Bangkok.[167] Menadue argued that prior to this point, Asia was just something that Qantas executives and passengers would rather "fly over" rather than integrate into a business strategy. "Historically," he told me, "Australia has seen itself as an outpost of empire, first the British Empire and now the American Empire. So, businesses, executives, and travelers have seen Asia as something that you fly over. A place you fly over and if you have to stop a few times, Singapore and Bahrain, it's unfortunate. But that's the way you get to Europe."[168] The archival paucity of strategic analysis in Qantas files at the National Archives of Australia regarding route development supports such a claim. Furthermore, Hudson Fysh's allegiance to the Commonwealth and shared services with British Overseas Airways Company on the Kangaroo Route after 1947, substantiate the idea that stopovers in Asia were a function of technological constraints rather than efforts to spur Asia-centric models of decolonisation.

To the consternation of some board members, Menadue wasted no time in tilting Qantas' business model towards East Asia. "I had to tell them that it was how we made our profits on the way to Japan" he told me. "The Kangaroo Route was a marginal earner. So, I think it probably was the first attempt for Australia to break through seriously into the Asian market."[169] Extant business intelligence suggested that most growth in the commercial aviation sector would take place in Asia well into the new millennium. Aside from being sound practice, it also reflected the best ways to practice reciprocity. "Qantas has a vested and rightful interest in obtaining fair opportunities in Asian markets," he told one Australian audience during a speaking tour. "We simply want the opportunities in those markets that Asian car-

---

167 In 1989 Qantas built on regular service to Bangkok by routing daily flights between Sydney and London through Bangkok's new international terminal. While this snubbed Singapore of regular Kangaroo Route service, the route "would cut two hours off the flight time and recognized passenger preferences for fast, long-haul flights and the growing appeal of Bangkok as a stopover in Asia." (*Canberra Times*, "One Stop too Long by Qantas," Thursday, May 19, 1988, page 13).
168 Ibid.
169 Ibid.

riers have in our traditional markets."[170] Capitalising on Qantas's all 747 fleet, Menadue increased the number of flights to Japan from four to twenty-four weekly during his three-year tenure with the company. By and large, the expansion of destinations increased the number of cities served in Japan. Japan Airlines reciprocated with additional flights to Sydney. At the same time, Menadue cultivated nominal interest in Qantas service to South Korea, provided the South Korean government allow its citizens to travel to Australia, as well as offer reciprocal access to its commercial aviation market. Future markets would also include China and Taiwan.[171] Menadue carefully guarded the Australian market from airlines hoping to poach Qantas passengers— noting at one point that upwards of twenty airlines operated flights on some portion of that corridor—who traditionally traveled to Asian and European routes through Singapore. Menadue championed careful management of "open skies" in Australian air markets, where carriers such as Sri Lankan Airlines and Korean Airlines aspired to run their own hubbing operations.[172] While this might have been successful in the short term, in the long run, the rise of Persian Gulf carriers including Emirates Airlines, Etihad Airlines, and Qatar Airways developed a pipeline to European destinations through hubbing operations at Dubai, Abu Dhabi and Doha respectively.

---

170 John Menadue, "An Address by John Menadue, Chief Executive, Qantas Airways Limited to Guild of Business Travel Agents, Brisbane, May 22, 1989," John Menadue Papers, MS Acc 11.054, Box 8, National Library of Australia, Binder: QF Speeches, 1988/1989.

171 On China, see John Menadue, "A Speech by John Menadue, Chief Executive, Qantas Airways Limited to Australia-China Chamber of Commerce and Industry, Brisbane, August 26, 1988." In another speech he showed the range of his understanding of the complexities of navigating sanctions and social turbulence. In the wake of the Tiananmen Square massacre in June 1989, he told a Melbourne audience, "One lesson of Tiananmen Square is the unpredictability of events, whether they be within nations or between nations. It has been a reminder, a very powerful reminder, of the need to strive for knowledge, understanding, and when its most needed, compassion. This is just as important amongst friends like Australia and Japan as it is amongst any two nations serious about gaining the greatest benefits from a re4lationship. It is vital that after the events in China, we do not turn our backs on that country. We live in the same global neighborhood. Australia's economic prosperity and well-being are inexorably linked to that of our neighbors including China." See John Menadue, "An Address by John Menadue, Chief Executive, Qantas Airways Limited, to Australia-Japan Society of Victoria, Melbourne, June 23, 1989," John Menadue Papers, MS Acc 11.054, Box 8, National Library of Australia, Binder: QF Speeches, 1988/1989.

172 In an address to the Guild of Business Travel Agents in Brisbane, Menadue noted, "Qantas is but one of more than 20 airlines providing services between the U.K. and Australia—the market is far from a Qantas-British Airways duopoly." John Menadue, "An Address by John Menadue, Chief Executive, Qantas Airways Limited to Guild of Business Travel Agents, Brisbane, May 22, 1989," John Menadue Papers, MS Acc 11.054, Box 8, National Library of Australia, Binder: QF Speeches, 1988/1989.

Menadue did not sidestep Australia's historical anxiety towards Asia and Asians. Citing a then-current survey, Menadue acknowledged that "70 percent of Australians favoured an expansion of Japanese tourism—but 75 per cent were not in favour of any increased Japanese investment." To counter such sentiment, Menadue played to the better angels of the Australian business community. "Frankly," he concluded, "if we faced the realities in this capital-hungry country, and adopted a more positive, more pragmatic view, we would seize the opportunities for growth and prosperity offered by foreign investment—notably at this time, Japanese investment."[173]

Similarly, he had to keep customers and shareholders confident that he was not abandoning the Anglosphere, though he knew that the traditional Kangaroo Route tied up planes in ways that were not productive with long-stretches on the ground rather than in the air. In one speech, after laying out the case for Asian integration, he circled back, noting (emphatically), "But this does not mean Qantas *has lost sight of other markets, least of all the U.K. traditional ties remains strong.*" This, even though the Asian routes returned the airline to profitability.[174]

Menadue personally lobbied government officials, business associations, and civic clubs to elevate Asian linguistic and cultural competencies in Australia. He also pioneered a six-million-dollar investment programme in Asian cultural land language studies.[175] Institutionally, Qantas entered a tie-up with the Ministry of Education to participate in the "Languages of Australia" initiative. Other innovative approaches included a $600,000 program at the university level for an Asian Language Cadetship Programme that provided in-country language training in Asia as well as cultural instruction for students in the state of Victoria. These initiatives not only bolstered the quality of Qantas' customer service but also contributed to Australian engagement with Asia generally. By the end of 1988, twenty-five percent of cabin crew possessed some Japanese language proficiency. Another 350 employees

---

173 John Menadue, "An Address by John Menadue, Chief Executive, Qantas Airways Limited, to Australia-Japan Society of Victoria, Melbourne, June 23, 1989," John Menadue Papers, MS Acc 11.054, Box 8, National Library of Australia, Binder: QF Speeches, 1988/1989.

174 John Menadue, "An Address by John Menadue, Chief Executive, Qantas Airways Limited to Guild of Business Travel Agents, Brisbane, May 22, 1989," John Menadue Papers, MS Acc 11.054, Box 8, National Library of Australia, Binder: QF Speeches, 1988/1989. On Menadue's departure, see Tom Burton, Deborah Light and Peter Shark, "Dogfight in the Qantas Cockpit," *Sydney Morning Herald*, Saturday, July 22, 1989, clipping in MS Acc 11.054, Box 8, ephemera, National Library of Australia.

175 John Menadue, "An Address by John Menadue, Chief Executive, Qantas Airways Limited, to Australia-Japan Society of Victoria, Melbourne, June 23, 1989," John Menadue Papers, MS Acc 11.054, Box 8, National Library of Australia, Binder: QF Speeches, 1988/1989.

engaged in Japanese language training. By 1991, the airline hoped to retain 700 employees with Japanese skills.[176]

As had a long line of industry insiders before him, Menadue also pointed to connections between international tourism and the lacklustre state of airport and hotel infrastructure in Australia. Speaking globally, Menadue noted, Australia seemed to be trending towards attitudes prevailing towards airport infrastructure in the Anglosphere: "Like Europe and the United States, Australia is suffering from deficiencies in the infrastructure and support services—notably airport terminals."[177] "We have 11 gateways for 16 million people or one for every 1.4 million people," he noted, "yet 85 per cent of visitors use only three of these airports; Sydney, Melbourne and Brisbane all of which suffer from inadequate facilities for international passengers."[178] His frame of reference was clear. "Competing destinations such as Singapore and Bangkok offer excellent arrival and departure facilities." In contrast, he then explained, "The Sydney terminal—despite the modest expansion now underway—will not adequately meet demand in the foreseeable future."[179] Protecting the strategic advantage of its gateway traffic should not be taken for granted, he warned:

> Sydney's worldwide reputation as the focal point for tourism in Australia is a priceless asset for [New South Wales], and both the industry and government are recognising that they must jointly protect their investment. As the main market with the most traffic, Sydney naturally attracts the most direct air services. Concentration of air service attracts business traffic and influences international companies such as banks in locating their Australian headquarters.
>
> Gateway status also creates the visitor volume which [New South Wales] needs to compete for tourist spending against other states like Queensland whose well-established resorts attract far longer stays. The other states want your business and will quickly take advantage of any complacency in NSW. So will our competitors in other countries and that will be Australia's loss.[180]

---

176 Ibid.

177 John Menadue, "An Address by John Menadue, Chief Executive, Qantas Airways Limited to Guild of Business Travel Agents, Brisbane, May 22, 1989," John Menadue Papers, MS Acc 11.054, Box 8, National Library of Australia, Binder: QF Speeches, 1988/1989.

178 John Menadue, "An Address by John Menadue, Chief Executive, Qantas Airways Limited, 'Managing the Growth of Inbound Tourism' to Committee for Economic Development of Australia, Sydney, April 26, 1988," John Menadue Papers, MS Acc 11.054, Box 8, National Library of Australia, Binder #2: Qantas Speeches, 1986–1988.

179 John Menadue, "An Address by John Menadue, Chief Executive, Qantas Airways Limited, to a Joint Meeting of the Rotary Club of Newcastle and The Newcastle Chamber of Commerce and Industry, Newcastle, November 9, 1987," John Menadue Papers, MS Acc 11.054, Box 8, National Library of Australia, Binder #2: Qantas Speeches, 1986–1988.

180 John Menadue, "An Address by John Menadue, Chief Executive, Qantas Airways Limited, to the Eastern Australian Chapter of the Pacific Travel Association (PATA), Sydney, February 18,

Eventually, Menadue and Qantas took measures into their own hands, pouring money into "check-in facilities and our Captain Club Lounges."[181] With reference to hotels, the CEO cited a deficiency across category types, beginning with five-star hotels in Sydney and continuing down through modest three-star offerings. Anticipating construction of Sydney's Darling Harbour mega-complex of hotels, he told a Pacific Area Travel Association meeting, "if Baltimore [Maryland, in the United States] can effectively absorb 5000 new hotel rooms through a harbour-side facelift, surely we could pack at least that many into some redevelopment of the top tourist city in the South Pacific."[182]

Menadue's efforts bore immediate and sustainable fruit. By late 1989, he projected that Qantas, Japan Airlines, and ANA would operate forty direct flights between the two countries. Already, he observed, "Japan has overtaken the UK and Europe in the Qantas schedules with 50 per cent more flights per week planned by November [1989] (24 against 16)."[183] More globally, Menadue noted at the 1989 World Aerospace Summit in Davos, Switzerland, "Qantas' summer schedules list 16 flights a week from the UK and Continent, 22 from the US and Canada, and 55 a week from Asia and Japan." Even with a return to profitability during his tenure, differences with Qantas Board of Directors led to Menadue's departure from the airline in late 1989. Fundamentally, however, Menadue had altered the identity of Qantas in a way not replicated since. "In Australia," he wrote, "we are taking part in this growth, not as a small European nation (with a present population of only 16 million) on the southern edge of Asia, but as a proud Asia-Pacific nation with a strong European heritage and connections, adding to the diversity and richness of our region."[184]

1987," John Menadue Papers, MS Acc 11.054, Box 8, National Library of Australia, Binder #2: Qantas Speeches, 1986–1988.

181 John Menadue, "An Address by John Menadue, Chief Executive, Qantas Airways Limited to Guild of Business Travel Agents, Brisbane, May 22, 1989," John Menadue Papers, MS Acc 11.054, Box 8, National Library of Australia, Binder: QF Speeches, 1988/1989.

182 Ibid.

183 John Menadue, "An Address by John Menadue, Chief Executive, Qantas Airways Limited, to Australia-Japan Society of Victoria, Melbourne, June 23, 1989," John Menadue Papers, MS Acc 11.054, Box 8, National Library of Australia, Binder: QF Speeches, 1988/1989.

184 John Menadue, "'Aeropolitical Prospects for Australasia-Pacific,' A Paper by John Menadue, Chief Executive, Qantas Airways Limited, for *Leaders* Magazine, a Special Edition Devoted to Aviation," John Menadue Papers, MS Acc 11.054, Box 8, National Library of Australia, Binder #2: Qantas Speeches, 1986–1988.

West, Then East: Reprising Singapore as the Primary Entrepot

The disproportionate number of outward Australian journeys that require stopovers has shaped the nature of contact with Asia. Brevity and liminality characterise many representations of Australians' Asian encounter. This is evident in the way in which such contacts are described in the scholarly literature. Agnieszka Sobocinska acknowledged this in her study, *Australians in Asia*, writing, "Although Asian stopovers were short – usually only a few days and occasionally even a few hours – they tended to evoke strong responses."[185] While this makes no claim about the quality of such encounters, it identifies a feature in outbound travel from Australia that made stopovers an almost intrinsic element of overseas travel. These contacts preceded the age of aviation but established key hubbing points that also served for cross-cultural interaction. Of the age of sea travel, Sobocinska notes: "Travellers extrapolated from brief exchanges with servants and hawkers in Batavia, Singapore or Colombo to make pronouncements about the 'Orient' or 'natives' in general. It is common for travellers to extrapolate outwards from fleeting personal experiences; this is the basis of the authority held by explorers, travellers and travel writers."[186] And, while many Australian-Asian encounters have stretched over years or been deeply intimate, the sense of Sobocinska's work is that these contacts are fleeting more often than not. In the conclusion of the book she writes, "This book testifies that Australians *have* been here before; indeed, they have been visiting their neighbors for well over a century. Some visitors touched down for only a few hours, but other Australians have lived their entire lives in-between the two continents."[187]

The liberalisation of commercial aviation not only pulled Australian tourists towards Asia, including Japan, but also a broader array of entrepots on European travel routes. With the opening of the new Bangkok International Airport (Suvarnabhumi), Thai officials hoped to lure transit and stopover passengers away from Hong Kong and Singapore, though not without a fight. In a 2005 "Australia Day" speech, Murray Cobban, Consul General of Australia, announced, "Australians travelling on the 'Kangaroo Route' to London will have the option to stopover in Hong Kong, bringing more Australian tourists into the territory. 2004 also saw the inauguration of Virgin Atlantic flights to Sydney, while Hong Kong's Dragonair is expected to take up the route later this year."[188] Beijing's emergence as the most

---

185 Sobocinska, *Visiting the Neighbors*, 18.
186 Ibid., 26.
187 Ibid., 214.
188 Murray Cobban, "Australia Day 2005," Media: Speech 20050126, https://hongkong.china.embassy.gov.au/hkng/SP_20050126.html, accessed July 22, 2022.

global gateway to the Asia Pacific also provided an attractive destination or stop-over, with 777,000 Australians visiting mainland China in 2013 (with 719,000 Chinese returning the favor to the most southern continent).[189]

Finally, Gulf Coast carriers Etihad, Qatar, and especially Dubai-based Emirates, swooped up more Europe-bound traffic in Australia, to the consternation of Qantas and Singapore Airlines. Hopping over or eliminating Singapore as a stop-over was not a novel aspiration. As early as 1957, Qantas received approval to transit Europe via the United States in case of conflicts in the Middle East. Three decades later, an unnamed Australian airline threatened to launch "cheap direct flights to Europe that will omit the island as a stopover." Such a move, the Singaporean based *Times Straits* newspaper noted, "could cost Singapore about $31 million [Australian dollars] a year."[190] It would not be until the mid 2010s that Emirates launched a code sharing agreement with Qantas Airlines to circumvent Singapore. Proposed in 2012, implemented in 2013, and rescinded in 2017, the bold stroke revealed not only what Australians preferred in a stopover destination, but the nature of their encounters with Asia in general.

Qantas and Emirates touted the economies of scale they could achieve by routing Qantas flights to Europe (including a high frequency route to London) through Dubai versus Singapore. The agreement relieved Qantas' loss-making European routes, as well as shaved time off the Sydney to London route. Furthermore, the tie-up eliminated the need for a fifteen-year agreement between British Airways and Qantas on the Kangaroo Route, a move that freed up British Airways and Qantas to use premium long-haul aircraft on a greater number of Asian flights.

Despite these advantages, Singaporeans warned Qantas to not by-pass "the Lion City." There were too many reasons to keep Qantas' traditional stopover. Two of Qantas' Jetstar low-cost carrier subsidiaries flew substantial numbers of flights each week from Changi. Qantas had recently poured ten million dollars into its club lounge there, not to mention continued to share office space with British Airways landside.[191]

Not lost in these inter-airline machinations were the consequences for battling hubs in the broader global aviation picture. A critic of the Emirates-Qantas agreement observed, "Dubai is a key Changi rival and, Singapore, in a bid to ensure that the airport's premium hub status is not eroded, may decide to let Qantas have what it wants rather than risk a tie-up with Emirates potentially boosting Dubai's busi-

189 Nina Karnikowski, "Australian Tourists Flock to China," *Examiner*, https://www.examiner.com.au/story/1973209/australian-tourists-flock-to-china/, accessed July 22, 2022.
190 *Canberra Times*, "Singapore Over Cut Stopover," December 12, 1978, 9.
191 See Karamjit Kaur, "Qantas Should not Jilt Changi for Dubai," *The Straits Times*, July 30, 2012, Factiva.

ness."[192] The hubs also leveraged tourism. "In the first quarter [of 2013]," the *Straits* correspondent continued, "[Singapore] attracted 230,000 visitors from Australia – the fourth highest behind Indonesia, China, and Malaysia." But the Emirates tie up was only the tip of the iceberg in the growing competition between airlines and hubs for tourists. Greater flexibility not only from Middle Eastern carriers, but Chinese as well, altered the stopover playing field in the mid-2010s. "Exclusive Australian travellers are increasingly heading to Europe via the Middle East and China," Lisa Allen, travel writer for the *Australian*, wrote in the wake of the Emirates-Qantas tie-up in 2015.[193] Tourist agencies sensed the profitability of luring travelers, if even for a night, to destinations beyond Singapore and Hong Kong. The scale of the shift was significant. Allen reported, "Just over one million Australians travelled to Dubai on Emirates in the year to June 2014, while Etihad carried more than 300,000 passengers to Abu Dhabi. A further 170,615 Australians [passed through] . . . Qatar via Qatar Airways."[194] Australian travel agents were bullish on the trend, if for no other reason than the diversion that if offered travelers. "'People are attracted to the sand dunes of the Middle East and once the new cultural infrastructure projects open in Abu Dhabi, they will become an even hotter non-traditional destination for Aussies,'" opined travel expert David Goldman.[195]

Ultimately, however, Qantas' five-year experiment with Emirates did not consider perhaps two of the most important factors in flying: circadian rhythms and cultural preferences. As early as 2013, *Business Traveller* magazine proclaimed that Changi was still superior to the Middle East airports. "[The] Gulf airports may not offer the same comfortable transfers as, say Singapore Changi, which has won the Best Airport in the World award in our annual readers' poll for 25 consecutive years."[196] The magazine also opined that the fourteen-and-a-half hour flight to Dubai, versus the more circadian friendly eight-hour-flight from Sydney to Singapore, arriving just before bedtime, might not jive with many fliers.[197] Qantas discontinued the arrangement in 2017, removing Dubai from its network while continuing with a codeshare agreement with Emirates Airlines.[198] Dubai's loss was a win for Singapore and Changi, as "'A lot of economies are built

---

192 Ibid.
193 Lisa Allen, "Adventurous Fuel China, Middle East Short Stays," *The Australian*, January 17, 2015, Factiva.
194 Ibid.
195 Ibid.
196 *Business Traveller*, "Kangaroo Moves," February 26, 2013, Factiva.
197 Ibid.
198 Patrick Hatch, "Airport Transfers Could be on the Way Out," *The Age*, September 9, 2017, Factiva.

around those airports. Yes, they want you to connect there, but ideally they want you to spend a couple of days there. They're trying to grow their tourism, and both are very important trade hubs.'"[199] Midway into the Emirates-Qantas agreement, at least one travel agent found that for Australians transiting the Gulf, "finding quality hotels in some Middle Eastern destinations was . . . tough."[200]

John Menadue concurred with the argument that Australians preferred stopovers in the former East and Southeast Asian British city-states above others. This came down to questions related to language, efficiency, and amenities. "I'm certain that Singapore was an important entrepot and particularly Hong Kong, the city-states as I called them."[201]

> It was just another feature of the fact that Australian passengers and Qantas executives were much more at ease dealing with English speakers in Singapore and also in Hong Kong. It was just natural, it's just the way you do things rather than in other Asian countries. And, of course, the factor was Singapore and Hong Kong were city-states with small populations. They did have good technical support, maintenance, facilities at their airports, so that Singapore and Hong Kong became important cogs particularly in the Qantas traffic to Europe, but also through Hong Kong to other parts of Asia. But Singapore was the key entrepot for Qantas traffic on the way to Europe.[202]

While Bangkok relieved some pressure from Qantas' reliance on Singapore as primary entrepot, its urban congestion, among other factors, made it less attractive to passengers. "It never had, as a tourist destination," Menadue noted, "quite apart from the airport, the efficiency or attractiveness of Singapore for sheer efficiency, good food and so on."[203] Menadue was equally skeptical that the cities where the Gulf carriers were based could match the appeal of Singapore and Hong Kong. "I think . . . going through Qatar, Dubai . . . [it] just didn't compare with the efficiency, cleanliness and friendliness of Singapore and so Qantas withdrew from Doha and the Emirates."[204]

Aviation researchers corroborated *Business Travellers'* impressionistic statements. Steven Pike and Filareti Kotsi conducted several studies of travellers of different nationalities passing through various hubs to assess the likelihood of return visits. This included interviewing nearly five hundred Australians transiting Dubai. Of the twenty-one attributes Australians identified as most important for a stopover, the

---

199 Ibid.
200 Allen, "Adventurous Fuel China, Middle East Short Stays."
201 Menadue, interview with the author, in author's possession.
202 Ibid.
203 Ibid.
204 Ibid.

top twelve included: an interesting culture, comfortable flight times, ease of transit, entertainment and nightlife, good food, lots to see/do, good hotels, friendly people, new experiences, respectful attitudes towards women, interesting architecture, and previous familiarity.[205] While these were the most frequently identified characteristics of an ideal stopover, the importance of attributes differed significantly. They were, in order: safety, respect towards women, good hotels, friendly people, comfortable flight time, favorable flight schedules, ease of access, cleanliness, affordability, great food, favorable weather, opportunities for new experiences, lots to see/do, and English language use.[206] When asked to choose between preferred stopover destinations, Australians responded in descending order: Singapore, Dubai, Hong Kong, Bangkok, USA/Canada, Abu Dhabi, Kuala Lumpur, Indonesia, China, United Arab Emirates/Qatar, Japan, and India.[207]

Sobocinska's research tends to square with this preference for former British colonial city states stopovers. This was a consistent feature of historical contacts between Australia and the near north, as "After settlement, British colonial networks linked Australia to ports at Singapore, Hong Kong, [Sri Lanka] and India."[208] Of more recent Australian passengers, she wrote, "Tourists to Hong Kong or Singapore may not have been so obviously regulated, but many found it just as difficult to access the 'real' Asia. Many visitors were unwilling to venture too far into the unknown, and the comfort of their hotels made the heat and confusion outside seem less appealing."[209] The special role of "the business trip" in Australian commerce stretched the breadth of destinations visited by Australian travelers, but ultimately encompassed the well-trodden paths of *adjacent* empire. Language, personal security, and the consumer experience all favored the former city states (as well as Japan, minus the facility in English). Sobocinska notes of their appeal:

> Singapore, Hong Kong and Japan exerted a steady attraction for tourists eager for sightseeing and shopping [from the 1970s]. Singapore's tourism promotions promised the exoticism of Asia in a relatively safe, clean and Western environment and discounted package holidays to Singapore were astoundingly popular in the early 1970s. Hong Kong promised a mix of East and West: 'the Chinese younger set are as 'mod' and 'a go-go' as the youngsters in

---

205 Steven Pike and Filareti Kotsi, "Stopover Destination Image—Perceptions of Dubai, United Arab Emirates, among French and Australian Travellers," *Journal of Travel and Tourism Marketing*, 35:9 (2018), 1165.
206 Ibid., 1167.
207 Ibid., 1167.
208 Sobocinska, *Visiting the Neighbors*, 29.
209 Ibid., 94.

London, New York, San Francisco and Sydney' . . . Both destinations balanced the exoticism of the Orient with references to their British colonial heritage and luxurious hotels, offering a frisson of Oriental excitement with the assurance of safety and security.[210]

The idea of an essential divergence of cultural preferences between what is considered East Asia and West Asia (or the Persian Gulf) is part of a larger debate concerning the coherence or compartmentalisation of regions within the broader Indo-Pacific region. *Singapore Nation* raised this issue in an article entitled, "Retaking West Asia from the Middle East." Although the author makes some cultural assumptions about bias and discrimination, his core thesis of the unity and disunity of the Indo-Pacific bears consideration. The columnist wrote, "Given East Asia's abhorrence of disorder and our broad preference to keep religion largely a private matter, most of us also tend to think of the Gulf states and their key neighbors as a zone of endless troubles, a headache best avoided when considering investments and certainly travel, unless you are a Muslim and planning a pilgrimage."[211] They then address the consequences of such cultural myopia: "Indeed, it is a pity that we routinely refer to it as the Middle East, when, in truth, the key states, Israel included, ought really to be approached as West Asian nations. By not doing that, we are missing out on closer awareness of a thriving neighborhood with some idea of where it is going and how it will get there."[212] While this has little to do with Australia and the travel preferences of its citizens, it calls into question how the vast region of the Indo-Pacific is culturally coherent or heterogeneous and may explain, at least in part, the stopover preferences of different individuals throughout the region when considering Asia as a cultural construct.

Conclusion: Australians in the Asian Entrepot

Australians' dependence on ultra-long haul air travel, stopovers, and airports reflects their geographic position as the ultimate point of origin. A keen interest in the passenger amenities of airports, including Berlin's Tempelhof, Amsterdam's Schiphol—home of Imperial Airways chief competitor in the era of boat planes, KLM—, and Commonwealth heavyweights Croydon and Heathrow, reflected as much a desire to improve airports in Australia as it did in finding the most convenient way to travel from one end of the earth to another. Singapore emerged as the first Asian entrepot of significance during the early age of aviation, given its

---

210 Ibid., 170.
211 *The Nation* (Singapore), "Retaking West Asia from the Middle East," March 11, 2017, Factiva.
212 Ibid.

central position on the Europe to Australia/Batavia routes. It was likely during those years that Australians developed a preference for stopovers in locations of salubrious weather, high levels of service and courtesy, English-speaking interactions, and ample shopping opportunities. Hong Kong emerged as an alternative, but only after advances in aviation technology made the journey comparable in length to that of the Singapore route. Hong Kong's position, however, as an enduring entrepot, was never as secure as that of Singapore, given the uneasy relationship between the crown colony and the agreements for return of the territories to mainland China in 1997.

Eventually, the boom in twenty-first century aviation in the Indo-Pacific cost Qantas market share with the incursions not only of American carriers, in the immediate wake of World War II, but also the Middle Eastern/West Asian carriers Emirates, Airlines Etihad Airlines, and Qatar Airways. Australian encounters in a wider ambit across Asia not only encouraged Australians' exploration of Southeast Asian destinations such as Bali, but also manifest cultural affinities with stopover points like Singapore, which enjoyed cultural advantages in the minds of Australians to alternatives, namely in West Asia.

The hubbing phenomenon would not go unchallenged, however, what with the articulation of a new model of air travel: low-cost carriers who engaged in point-to-point routes using new aircraft, including updated versions of the Boeing 737, the Airbus 320 series, and the Boeing Dreamliner. What Southwest Airlines pioneered in the United States, and Ryanair and EasyJet applied in the European Union, required some adaptation in Asia, but caught on and theoretically threatened the staying power of hubs. Would a new mode of travel, engaging a whole new "republic" of working- and middle-class passengers, kill the hub?

Chapter 5
Low-Cost Carriers and Hubs in the Indo-Pacific

Being a big low-cost carrier globally demands charting a new path.
– Surateet Das Gupta[1]

We are . . . offering an affordable and alternative version of the 'Kangaroo Route' all the way from London via Kuala Lumpur to Melbourne, Perth and the Coast in Australia.
– Azran Osman-Rani, CEO of Air Asia X[2]

Mr. Fenandes has complained long, hard and frequently about KLIA2, the charges for using it and its Taj Mahal-like nature.
– Corporate Travel Community[3]

Introduction

The strategy of hubbing for tourists—i.e., optimising airport infrastructure, launching hometown long-haul airlines, and relentlessly promoting over-the-top hotels—flourished in the Indo-Pacific region in the second half of the twentieth century and beyond. The approach, however, was not without challengers. There were many moving pieces in the aviation sector that potentially threatened the hub-and-spoke model of global city ascendance, particularly with the advent of low-cost-carriers.

Low-cost-carriers emerged in the United States and Europe in the wake of de-regulation. Southwest Airlines, using the expansive state of Texas as its base, rapidly spread its wings on a point-to-point network model that took advantage of a single aircraft type (Boeing 737's), secondary airports, rapid turnaround times, and no-frills inflight services, to challenge legacy carriers for market dominance in the United States. Similarly, Ryanair and EasyJet vied for market share amongst the legacy carriers in Europe, contributing, in part, to the consolidation of national carriers, including KLM and Air France. In both the United States and Western Europe,

1 Surajeet Das Gupta, "IndiGo on long-haul flight path, to take final call on buying Air India," *Business Standard*, February 12, 2018, Factiva.
2 *United News of India*, "Air Asia X to Increase Frequency on KL-Stansted-London Route," April 21, 2008, Factiva.
3 Corporate Travel Community, "AirAsia's Tony Fernandes Again Laments the Lack of Low-Cost Terminals in Malaysia; Points to Japan for Inspiration," https://corporatetravelcommunity.com/analysis/airasias-tony-fernandes-again-laments-the-lack-of-low-cost-terminals-in-malaysia-points-to-japan-for-585492, accessed January 17, 2023.

https://doi.org/10.1515/9783111326641-006

low-cost-carriers benefitted from a single regulatory market, in one case the fifty states in North America and in the other the European Union.[4]

Low-cost-carriers caught on much later in Asia. Nevertheless, they had an equally disruptive effect on the bottom lines of full-service legacy carriers. Although their respective operations often crossed international boundaries, their devastating effect on legacy carriers in Southeast Asia's domestic markets forced traditional full-service airlines to either join the low-cost market or focus more intently on high-quality service and long-haul markets.

Ultimately, however, what impact did the low-cost-carrier phenomenon have on the "hubbing for tourists" model? While there was much talk of point-to-point service challenging the primacy of the hubbing model in places like Dubai, Singapore, and Hong Kong, low-cost carriers effectively *bolstered* hubbing practices throughout the Indo-Pacific. While low-cost-carriers generally sought out secondary airports in Europe and the United States, such facilities were not readily available in a region rapidly rolling out its first generation of hubs.[5] If there were exceptions, such as in Bangkok, most Indo-Pacific airports catered to *both* full service and low-cost-carriers. Put another way, Asian low-cost-carriers typically conformed to the marketing preferences of a) airport administrators, who were eager for profits beyond landing fees from their hubs; and b) the acquisitive posture of their passengers, who were all too willing to empty the savings they pocketed from cheaper plane tickets on consumer goods available in shopping centre-like hubs at Singapore, Kuala Lumpur, and Hong Kong.

Historiography

An emerging field of scholarship on low-cost-carriers assesses the most transformative aspect of commercial aviation since deregulation. Several important studies illustrate the commonalities between airlines that have entered this volatile sector of civil aviation, as well as the ways in which the budget model differs by

---

4 Shelley Vishwajeet provides a helpful overview of the history of LCCs in *The IndiGo Story: Inside the Upstart that Redefined Indian Aviation* (New Delhi: Rupa Publications India Pvt Ltd., 2018), 23–38.

5 Furthermore, as LCCs have consolidated their positions in American and European markets, many have shifted their base of operations from LCCTs to hubs. See Frederic Dobroszkes, Moshe Givone, and Timothy Vowles, "Hello Major Airports, Goodbye Regional Airports? Recent Changes in European and US Low-Cost Airline Airport Choice," *Journal of Air Transport Management*, volume 59, March 2017, 50–62.

global regions. Sven Gross and Michael Lück's edited collection, *The Low-Cost Carrier Worldwide*, breaks Asia and Oceania into three separate chapters—one on Asia and the Pacific; a second on India; and a third on Australia and New Zealand. In the first, Semisi Taumoepeau emphasizes the dynamism of the LCC model in the Asia Pacific region, as well as the key hubbing functions of Sydney, Auckland, and Melbourne for LCCs in the South Pacific;[6] in the second, Tarun Shukla problematises the role of low cost carrier airports in an Indian market without secondary airport infrastructure in cities such as Delhi;[7] and, third, Michael Lück and Sven Gross examine the ethical dilemma posed by a lack of secondary airports in Australia and New Zealand, which makes it less likely that "major airports will agree on [lower] concessions, given the increased demand by the LCCs and the lack of secondary airports in this region."[8]

Similarly, John Bowen's recent book, *Low Cost Carriers in Emerging Countries*, highlights several key moments in the evolution of recent aviation history in the Indo-Pacific region that have encouraged the emergence of LCC's, including the 1998 Asian Financial Crisis, which led to deregulation of civil aviation generally; the attack on the United States on September 11, 2001, which caused the contraction of fleet sizes, making a glut of aircraft available to new start-ups; the SARS crisis, which further hollowed out capacity on legacy carriers, opening opportunities for budget carriers; and, finally, the Great Financial Crisis of 2008, which also winnowed poorly capitalised legacy carriers from among the ranks of competitive commercial airlines.

Bowen's contribution to the study of low cost carriers in the Indo-Pacific, which includes a chapter on the emergence of low-cost-carriers in the shadow of megahubs in the Gulf; an overview of the burgeoning Southeast Asian market, where capacity and safety remain primary concerns; and, finally, trenchant analysis as to why LCC's have caught on in India while China remains largely indifferent to low cost carriers, provides unapparelled assessment of the variety of LCC's across the Indo-Pacific.[9]

Third, Max Hirsh's work on aviation infrastructure in Southeast Asia and China makes an important contribution to discussions about the nature of integrated city networks to cater to *all* of Asia's aspiring air passengers. While Hirsh's discussion of the (poor) level of integration at present goes beyond the scope of

---

6 See Semisi Taumoepeau, "Low Cost Carriers in the Asia Pacific," in Sven Gross and Michael Lück, eds., *The Low Cost Carrier Worldwide* (Abingdon, Oxon, UK: Routledge 2014), 113–138.

7 Tarun Shukla, "Low Cost Carriers in India," in Gross and Lück, eds., 139–154.

8 Lück and Gross, "Low Cost Carriers in Australia and New Zealand," in Gross and Lück, eds., *The Low Cost Carrier Worldwide*, 163.

9 John Bowen, *Low-Cost Carriers in Emerging Countries* (Amsterdam: Elsevier, 2019).

this book in looking at linkages between residential areas, central business districts, and hubs themselves, his work illuminates the importance of practitioners and academics talking to each other about making low-cost travel opportunities accessible for middle and working class Asians.[10] Such insights align with earlier work by Richard De Neufville, whose article, "Low-Cost Airports for Low-Cost Airlines: Flexible Design to Manage the Risks," provides an authoritative paradigm for designers to "recognize explicitly the uncertainties that threaten the planned developments and investment, and develop ways to respond easily to the many different scenarios that might develop."[11]

Finally, Urs Bingneli and Mathieu Weber offer a sanguine outlook for the survival of hub airports in the age of low-cost carriers. While it is true that low-cost-carriers have laid low the domestic dominance of legacy airlines, Binggeli and Weber both doubt the sustained viability of budget airlines on medium and especially long-haul routes, where a) turn-around times are decidedly longer than the twenty-five minutes targeted by LCCs; b) single cabin configurations drag down revenue needed to cover elevated expenses; and, c) the lack of in-flight frills on journeys longer than eight-hours challenge even the most ascetic traveler. All these factors, they argue, make it difficult for budget airlines to turn a profit or convince the flying public to sign up for such arduous journeys across continents. Given such constraints, they contend, it is unlikely that hubs will cease to dominate long-haul travel.[12] Likewise, by inference from the current state of the low-cost-carrier network configurations, short and medium haul traffic in Asia will continue to entrench the role of hubs throughout the region.

---

10 Max Hirsh, *Airport Urbanism: Infrastructure and Mobility in Asia* (Minneapolis: University of Minnesota Press, 2016).

11 Richard De Neufville, "Low-Cost Airports for Low-Cost Airlines: Flexible Design to Manage the Risks," *Transportation Planning and Technology*, 31:1, 53.

12 Urs Binggeli and Mathieu Weber, "A Short Life in Long Haul for Low-Cost Carriers, Travel, Trasport & Logistics, McKinsey&Company, https://www.mckinsey.com/~/media/McKinsey/Indus tries/Travel%20Transport%20and%20Logistics/Our%20Insights/A%20short%20life%20in%20long %20haul%20for%20low%20cost%20carriers/A_short_life_in_long_haul_for_low%20cost_carriers. pdf, accessed January 9, 2023.

The Asian Rise of the Low-Cost-Carriers

The long-term impact of low-cost-carriers in the Indo-Pacific began as soon as Malaysia and Singapore decided to split the routes and planes from their previously shared airline, Malaysian Airlines, in 1972.[13] When Singapore chose the long-haul routes, it opted for an infrastructure appropriate for global travel, something made manifest in Changi International Airport. An unforeseen benefit of Changi was the long-term possibility of expanding the airport by stages. The Malaysian government opted for the domestic routes. Most flights originated in Kuala Lumpur at Subang International Airport. In the process of developing new routes, Malaysian Airlines charted its own international network. This placed it in direct competition with Singapore Airlines. Focusing on domestic routes became increasingly difficult.

Early in September 2001, former music executive Tony Fernandes secured an interview with Malaysian Prime Minister Mahathir bin Mohamad to seek permission to operate a low-cost airline domestically. While Fernandes had dreamed of operating a long-haul airline since his days in prep school in Great Britain, his advisor, a partner in the launch of Ryanair, persuaded him to focus on short-haul routes with rapid turn-around times. This was a decision more conducive to the strategies that had made low-cost-carriers profitable in Europe and the United States. In the event, Fernandes breathed fire about competing against Singapore Airlines (an incongruous comparison between a start-up budget outfit and a full-service long-haul juggernaut), only to be met with an offer from the prime minister to take over Malaysian Airlines' exorbitantly priced domestic routes if he, Fernandes, could find an airline to service them. By securing AirAsia, an essentially defunct start-up with service between Kuala Lumpur and a Thai beach outpost, Fernandes began his assault on Malaysian Airlines' over-priced domestic routes, securing even more cities from the government in 2006.[14]

But what did this portend for the hubbing troika of luxury airports, long-haul airlines, and prestigious hotels? Early experience from the American and European markets suggested that operators would look for or request secondary airports where operational costs were typically lower than at traditional hubs. Kuala Lumpur was something of an exception, having recently completed a new international airport at the turn of the twenty-first century. Initially, AirAsia opted to base its flights out of the former Subang International Airport, close to

---

13 Ken Hickson, *Mr. SIA: Fly Past*, foreword by JY Pillay (Singapore: World Scientific Publishing, 2015), 62.
14 See Tony Fernandes, *Flying High: My Story from AirAsia to QPR* (London: Portfolio Penguin, 2017), 86–89.

the city Kuala Lumpur. This was convenient for cost-conscious passengers targeted by the new wave in flying. As noted in chapter 2, the new terminal was nearly forty miles from the city.

Eventually, however, competition between Singapore and Kuala Lumpur spurred the localised development of building barebones low-cost carrier terminals adjacent to the more opulent terminals. The Malaysian Airport Authority, sensing the crescendo of enthusiasm for AirAsia's low fares, constructed a basic low-cost carrier terminal two kilometers from the new Kuala Lumpur International Airport. Singaporean airport authorities quickly followed. These structures sufficed for a handful of years before national officials saw the opportunity to address two growing concerns: the declining likelihood that Kuala Lumpur's new airport would challenge Changi International for long-haul supremacy in Southeast Asia (discussed earlier in chapter 2) and the meteoric rise of Fernandes' AirAsia.[15]

To capitalize on the niche that low-cost-carriers offered at a central location in Southeast Asia, the Malaysian government constructed a second full-service terminal at KLIA, which opened on May 2, 2014.[16] The optimism of airport authorities and Fernandes's criticism towards the new structure illustrated the competing objectives of airport and airline management teams. On the one hand, airport authorities hoped to wring more profits out of an airport—which would include a shopping centre's worth of retail outlets—than simply landing fees and taxes.[17] Thus, at the terminal's opening, Puan Faizak Khairuddin, Senior General Manager of the Commercial Services Division noted:

> We are extremely pleased with this development. We have always maintained that the airport is more than just a place where you go to catch your flights. With KLIA2 we are able to articulate this proposition more clearly. With commercial space five times that of the old low-cost terminal, KLIA 2 is truly an airport in a mall. If anything KLIA2 is a destination in itself.[18]

---

15 See "The Shape of Things to Come; Are the New Low Cost Terminals in Asia-Pacific Pushing Ground Support Operations into the 21st Century?" https://www.aviationpros.com/home/article/10375865/the-shape-of-things-to-come, accessed June 22, 2023.

16 Michelle Teo, "Fast-Growing Low-Cost Carriers are now Paying for High-cost Airport Terminals," *The Edge Singapore*, May 26, 2014, Factiva.

17 For the position of airport authorities at KLIA2, see "Malaysia Airports Unveils Big Ambitions for new Low-Cost Terminal," https://www.moodiedavittreport.com/malaysia-airports-unveils-big-ambitions-for-new-low-cost-terminal/?format=pdf, accessed January 9, 2023.

18 "Malaysia Airports Celebrates KLIA2 Opening with Major Retail Campaign, https://www.moodiedavittreport.com/malaysia-airports-celebrates-klia2-opening-with-major-retail-campaign/?format=pdf, accessed January 9, 2023.

On the other hand, airline owners like Fernandes, whose bottom line hinged on the simplicity of airport infrastructure, carped at the need for a terminal with its own shopping mall and enclosed boarding ramps, a feature that raised rates for all airlines involved.[19] However, this did not stop Fernandes from building a four hundred room airport hotel attached directly to the new terminal.

As such, KLIA2, as the new terminal would be known, reinforced the hubbing model, a clear divergence from the alternative model of secondary airports to which budget carriers were accustomed throughout most of the rest of the world. First, KLIA2 was anything but a stripped-down facility. The fact that it became home to AirAsia flights, which by the time it opened accounted for a sixty percent share of domestic flights in Malaysia, did not mean that it mirrored the humbler facilities AirAsia used either at Subang or at the initial LCCT near KLIA.[20]

The addition of a major shopping complex to the airport suggested that airport authorities understood that passengers still had pent up discretionary income, even if they had opted to fly on a budget airline. Such administrative aspirations were consistent with the findings of Graham Francis, Alessandro Fidato, and Ian Humphreys, who note that at one European LCCT, passengers typically arrived one or two hours before departure. "Airport management realised that in order to maximize retail and concession revenue they needed to process passengers swiftly and create more opportunities for passengers to spend money."[21]

About the same time, officials at Singapore's Changi Airport shuttered its makeshift low-cost carrier terminal, opting for a hybrid solution. The Changi Airport Authority announced a new state of the art facility, Terminal 4, which showcased high-end retailers and amenities associated with Changi's world-class reputation, as well as automated check-in services that accommodated efficiencies championed by low-cost-carriers. Like Terminal 2 at KLIA, Changi's Terminal 4 was *not* a secondary, low-cost terminal. Changi officials slotted Cathay Pacific, a full-service Hong Kong-based legacy carrier that went head-to-head with Singapore Airlines, side-by-side with other full service and low-cost-carriers there. Second, unlike the previous LCCT at Changi, the new terminal boasted a full complement of restau-

---

19 For the AirAsia perspective, see Corporate Travel Community, "AirAsia's Tony Fernandes again Laments the Lack of Low-Cost Terminals in Malaysia; Points to Japan for Inspiration," September 23, 2019, https://corporatetravelcommunity.com/analysis/airasias-tony-fernandes-again-laments-the-lack-of-low-cost-terminals-in-malaysia-points-to-japan-for-585492, accessed January 9, 2023.

20 I've taken this percentage from AirAsia's presence at Kuala Lumpur International Airport as cited in a broader article about the phenomenon of LCC's in Joseph Noronha, "Expanding Across Asia," *SP Airbuz*, November 27, 2013, Factiva.

21 Graham Francis, Alessandro Fidato, and Ian Humphreys, "Airport-Airline Interaction: The Impact of Low-Cost Carriers on Two European Airports," *Journal of Air Transport Management* 9 (2003), 270.

rants and luxury retailers, including the duty-free offerings that in many instances served as "department stores" for travelers more accustomed to shopping enroute rather than at brick-and-mortar outlets.[22]

As LCC's expanded outside of AirAsia's home in Malaysia, a "federation of networks" belied other differences between full-service and budget airlines.[23] With well-placed financial and political partners in Indonesia and Thailand, AirAsia opened a clutch of subsidiaries, tenuously linked to the home base in Malaysia by flights, for example, between KLIA2 and Bangkok's Don Mueang International Airport. While the subsidiary model worked well for point-to-point travel across borders, it also illustrated limitations of an *international* network of LCCs. While Bangkok, Singapore, and Kuala Lumpur acted as key bases for AirAsia Thailand, Air Asia, and AirAsia X (a long-haul subsidiary of the original corporation), there was not necessarily any guarantee that flights on LCCs would be held for incoming and ongoing passengers.[24] While this was an aspiration, federated networks remained distinct from their inter-lined legacy counterparts.[25]

Hong Kong and Shanghai also reinforced the idea that budget airlines encouraged the hubbing model.[26] LCC's caught on rather late in Hong Kong. In the 1980s, British aviation authorities rejected low-cost forerunner Laker Airways' bid for a route to London from Kai Tak International Airport. Likewise, construction costs at

---

22 Belmont Lay, "Changi Airport Terminal 4 is Singapore's Newest Shopping Centre," May 5, 2017, https://mothership.sg/2017/05/changi-airport-terminal-4-is-spores-newest-shopping-centre/, accessed January 9, 2023. Tenants included Kate Spade, Coach, Swarovski, and Montblanc.

23 John Bowen should be credited for naming the subsidiaries of AirAsia "a flying confederation." I've given the same groupings the name of "federation of networks" because they do not truly represent an integrated, consolidated system like that of hub-and-spoke legacy carriers. See Bowen, *Low-Cost Carriers in Emerging Counties*, 163.

24 A correspondent for the *Channel NewsAsia* wrote a narrative of traveling by LCCs entitled, "Four Days, 11 Flights, 10,000KM: A Reporter's Journey to Examine the Growth of Asia's Budget Airlines," September 29, 2017, Factiva. As far as connectivity was concerned, he noted of one flight, "Tight connections on different airlines are a risk in the budget airline world, as many of the cheapest tickets offer no option other than to buy another ticket if a flight is missed. And buying at the last minute will likely mean paying to dollar."

25 "Jetstar Airways Pty. Ltd., the budget unit of Australia's Qantas Airways Ltd. also offers connections to its passengers when it was feasible, said Alistair Hartley, Jetstar's head of network planning. "'We would still do it in a very low-cost manner.'" See Gaurav Raghuvanshi, "Asian Budget Carriers Look Afar," *The Wall Street Journal Europe*, July 15, 2014, Factiva.

26 In his article, "City Leads World in the Busiest Air Routes," *South China Morning Post*, May 7, 2018, Factiva, Danny Lee notes, "Data from industry trade group the International Air Transport Association showed the share of seats offered by budget carriers versus full-service carriers in Southeast Asia hit 52.6 per cent in 2016, while in East Asia the number was just 8.4 per cent. . . . Low-cost carriers made up just 10 per cent of flights out of Hong Kong, but a third in Singapore and 40 per cent in Jakarta."

Chek Lap Kok dictated prohibitive landing fees and taxes for many low-cost airlines.[27] Hong Kong's first LCC, long-haul pioneer Oasis Hong Kong Airways, launched flights to Vancouver and London in 2005, but could not absorb rising operating costs (a common theme for LCCs when they venture into the realm of long-haul operations), ceasing operations in 2008. Additionally, Cathay Pacific purportedly discouraged competition from low-cost-carriers, at least until its purchase of the HNA Group's budget airline, Hong Kong Express, in 2019. As one newspaper noted in March of the same year:

> Apart from Hong Kong Express, airlines looking to start frills-free operations in the city have been met with opposition. In 2015, Hong Kong rejected an application from Jetstar Hong Kong Airways Ltd., saying it didn't comply with legal requirements to operate such a business. Cathay [Pacific] also voiced its objection to the budget airline – a venture of Qantas Airways Ltd. And China Eastern Airlines Corp. that was backed by billionaire Stanley Ho's Shun Take Holdings Ltd.[28]

Finally, China's reluctance to embrace low-cost-carriers at the expense of what John Bowen denominates "the big three" (Air China, China Southern, and Eastern China Airlines), boded poorly for a budget bonanza in Hong Kong and Shanghai prior to the pandemic.[29] This does not mean, however, that Hong Kong has been unattractive to LCCs as a destination. Hong Kong, alongside Singapore, commonly found itself the first target for aspiring LCCs, including AirAsia X (a long-haul subsidiary of AirAsia), Cebu Pacific, and Singapore Airlines affiliate Scoot Airlines. Furthermore, Hong Kong's strategic location on the hinge between East and Southeast Asia has made it a popular destination for South Korean as well as Japanese LCCs (including the ANA subsidiary Peach Aviation).[30]

Nevertheless, industry experts surmised that Hong Kong could accommodate LCCs with the construction of an additional runway and terminal space. In the early 2000s, airport planners embarked on a master plan for 2030. This included

---

27 In her article, "Why is it so Hard to Find Cheap Air Tickets in Hong Kong?" *South China Morning News*, July 15, 2013, Factiva, Tiffany Ap writes: "the airport does not give any discount on landing fees to budget airlines, unlike, for example, Singapore. As a result, Hong Kong has lagged far behind cities such as Singapore and Kuala Lumpur. With no secondary airport, airlines are faced with expensive landing fees at Chek Lap Kok and competition for slots in tight."

28 *Macau Daily Times*, "Cathay Buys HK Express to Move into Budget Air Travel," March 28, 2019, Factiva.

29 Bowen, *Low-Cost Carriers in Emerging Countries*, 215, 218.

30 As of the time of writing (April 2023), LCCs play a crucial role in Hong Kong's return to hubbing status. On my day of departure from Chek Lap Kok, I counted two AirAsia, two Hong Kong Express, two Cebu Pacific, and one Scoot flight departing from Terminal One on a crowded morning.

construction of a second terminal, which opened in 2007. Although the facility did not include any gates, Terminal 2 served as a check-in space for low-cost carriers. The same master plan included a third runway project, which opened in 2022 for testing, and an adjacent concourse of fifty-seven gates connected to Terminal 2, slated to open in 2024. Integration of the facilities with the existing Terminal 1 suggests that Hong Kong, as well as Singapore and Kuala Lumpur, have opted for spaces that interchangeably cater to both LCCs and FSCs. As such, the enhanced Terminal 2 project also includes a jaw dropping six-story Sky City shopping centre, leisure park, hotel, and performance space adjacent to the airport.[31]

As far as the mainland is concerned, military primacy in Chinese airspace mitigated the impact that low-cost-carriers exerted on Chinese hubs. While civil air corridors are expanding in China, journalist James Fallows documented at length the degree to which military priorities trumped commercial aspirations.[32] In the case of the LCC featured in Bowen's analysis of the country, Shanghai's Spring Airlines, the carrier services *both* Shanghai's international (Pudong) and domestic (Hongqiao) hubs. This reinforces the idea that both full-service airports cater to legacy and budget airlines.[33] Hongqiao is only second to Pudong in size, aircraft capacity, and its domestic focus. Its consumer comforts include anything you might find in Singapore's Changi, including a Louis Vuitton store, among others. Ultimately, the policies of the Chinese government, which favor state-owned airlines, suggest that "secondary airports," like Guangzhou's Baiyun International Airport, were simply "biding their time" until their primary carriers, namely China Southern, allowed them to challenge Chek Lap Kok as primate hubs in the Greater Bay Area.

Southeast Asian countries Thailand, the Philippines, and Indonesia have gravitated towards the use of secondary airports, but not in the same sense as those in the United States and Western Europe. In Bangkok (hub for AirAsia's Thai subsidiary), Manila (a key hub for LCC Cebu Pacific) and Jakarta (hub for LCC Lion Air), the expansion of the flying public—caused primarily by the advent of LCCs in the 2000s—strained existing or newly constructed hub facilities. When Bangkok's new Suvarnabhumi Airport opened in the late 2000s, for example, plans to

---

31 Carine Chow, "Third Runway Set for Take-off by Mid-2022," *The Hong Kong Standard*, September 8, 2021, https://www.thestandard.com.hk/section-news/section/11/233995/Third-runway-set-for-take-off-by-mid-2022, accessed April 1, 2023. Also see Airport Technology, "*Hong Kong International Airport (HKIA) Expansion, Hong Kong*," September 1, 2022, https://www.airport-technology.com/projects/hong-kong-international-airport-hkia-expansion/, accessed April 10, 2023.
32 James Fallows, *China Airborne: The Test of China's Future* (New York City: Vintage, 2012).
33 Ibid., 218.

shutter Don Mueang International Airport were aborted, given insufficient capacity at the new airport for a surge in demand for slots. During the previous decade, Thai officials deregulated commercial aviation, effectively opening the skies to a spate of low-cost carriers, chief among them AirAsia Thailand. Officials temporised after the opening of Suvarnabhumi, vacillating on the fate of Don Mueang, eventually reopening it for use by both low cost and full-service carriers. Today the airport serves LCCs exclusively.

In the cases of Jakarta and Manila, aging airports designed in the 1970s and 1980s reached capacity. Airport starchitect Paul Andreu had designed the Jakarta airport to breathe with the flora of the Indonesian tropics, garnering an Aga Khan award for architecture in the process. "It is built," Andreu later reflected, "like the villages you fly over on arrival, by structures covered with tiles in the midst of a luxuriant garden and joined by open galleries that run the length of the garden to serve as passageways. The decision here was a result of a climatological and economic analysis and a reflection on Indonesian use of space."[34] In marked contrast, the evolution of Manila International Airport (later rechristened Ninoy Aquino International Airport), underwent a less than flattering rebuild in the 1980s before Paul Andreu designed an architecturally distinctive structure in the late 20th century. As an Asian Development Bank report noted of the new international terminal completed in 1983, Manila's declining gateway reputation accelerated even as capacity increased. A post construction project report in 1984 noted, "It is difficult to understand why the appearance of the surrounding of this international gateway is so neglected despite the fact that landscaping works and gardening service can be undertaken relatively cheaply."[35] The assessment continues:

> The airport Project had the expressed goal of providing facilities which would create a favorable impression on international travellers which in turn would reflect the development progress of the country. While the new facilities are an improvement over what was previously available, the visual impressions fall short of what experienced observers would expect after completion of such massive construction and a run-in period of two years. Sparse interior decoration, and unsophisticated lighting heighten these shortcomings.[36]

While Jakarta's Soekarno-Hatta International retained a semblance of Andreu's aesthetic distinction, the onslaught of budget carriers (namely Lion Air) raised

---

34 Paul Andreu, *J'ai Fait Beaucoup d'Aerogares . . .: Les dessins et les mots* (Paris: Descartes & Cie., 1998), 28, translation by the author.
35 Asian Development Bank Post Evaluation Office, "Project Performance Audit Report: Manila International Airport Development Project in the Republic of the Philippines," October 1984 (Manila: Asian Development Bank, 1984), 10.
36 Ibid., 11.

calls for further expansion, only a year after a third terminal opened in 2016.[37] Ironically, the Indonesian Transportation Ministry rejected Lion Air's bid to build its own LCC airport just a year before the inauguration of Terminal 3.[38] In the case of Manila, Ninoy Aquino International Airport continued a precipitous free fall in appeal and efficiency with the growth of Cebu Pacific (among other carriers).[39] As one journalist noted, "Philippine air terminals (in fact, all terminals of any kind) leave much to be desired in terms of amenities – especially toilets, art, efficiency, and the attitude of its personnel. Many of our terminals look like run-down government offices."[40] Clark International Airport, a repurposed United States military airfield near Manila, has attracted full service carriers (including Emirates) as well as LCCs for whom capacity at Ninoy Aquino International Airport is not sufficient. In all three cases, announcements have been made to the effect that new terminals will be built.

In Australasia, Australian and New Zealand hubs showed no sign of flagging in the age of low-cost-carriers. Sydney, Auckland, and Melbourne served as important hubs for tourists transiting South Pacific islands, Asia, and North America. As one scholar noted, the viability of low-cost carriers among the scattered South Pacific Islands hinges on the presence of hubs in New Zealand and Australia where passengers from intercontinental flights might be collected in numbers large enough to create the economies of scale LCCs require.[41] At present, Sydney's Kingsford Smith Airport encompasses separate international and domestic terminals. A new airport at Badgery's Creek, some distance from Sydney, will open around 2025. It will not have restrictive curfew hours, meaning that it could conceivably operate twenty-four hours a day. Without making a distinction between domestic and international flights (it is indeed too far away from Kingsford Smith to facilitate seamless transfers), it endangers the primacy of Sydney's historical hub, but could serve as an outlying facility for low-cost and full-service carriers alike.

While the fate of Sydney's aviation catchment remains in question, historical continuities along the Old Kangaroo Route persist, if only in the affinities shared

---

37 Bowen, 177.

38 *The Jakarta Post*, "Lebak Airport Megaproject Scrapped," November 17, 2015, Factiva.

39 Martin A. Khanser provides a thumbnail sketch of the origin story of Cebu Pacific in *John L. Gokongwei Jr.: The Path to Entrepreneurship* (Manila: Loyola Schools, Ateneo de Manila University, 2007), 89–90, 168–170.

40 Paulo Alcazaren, "Portal of the Jet Age," *The Philippine Star*, January 8, 2016, https://www.phil star.com/lifestyle/modern-living/2016/01/08/1540557/portal-jet-age, accessed April 1, 2023.

41 Semisi Tatumoepeau discusses the strategic role Australian hubs play in distributing tourists to South Pacific islands in "Suitability of the Low-Cost Airline Model in the South Pacific Region," in Duval, ed., *Air Transport in the Asia Pacific*, 93–112.

by Singapore and Australia. Not long after AirAsia had established itself in Thailand, Qantas launched its own federation of networks under an umbrella operation, Jetstar.[42] Distance made Australian (and subsequently its Singapore Airlines counterparts, Tiger Air and Scoot) low-cost operations unique. These carriers extended the international footprint of LCCs from between one to four hours flight time to between seven and ten hours flight time, linking Sydney and Melbourne with Singapore, Tokyo, Seoul, and Bali. With the inauguration of Singapore Airline's Scoot in 2012 (with SIA's Tiger Air integrated in 2016), two traditional powers on the Kangaroo Route made it known that they hoped to reclaim market share lost to Gulf carriers, who essentially gutted European capacity from SIA and Qantas, as well as curb the inroads on regional routes of upstart airlines from ASEAN countries.[43]

Farther west, Indian low-cost-carriers also accelerated hubbing developments in a market without secondary airports. Today's largest domestic carrier, IndiGo, based its model on the AirAsia strategy of using a single aircraft model in classless seating configurations, deployed internet booking to avoid a costly ticketing system, and executed rapid turnaround times with employees fulfilling multiple functions. In terms of infrastructure, budget airlines in India did not have secondary airports to exploit, a feature that would have further reduced costs associated with aerobridges and complex baggage systems. Instead, at the airlines' main base, Delhi's Indira Gandhi International Airport, IndiGo flights were scheduled for terminals two and three depending on whether they served international or domestic destinations.[44] A third terminal, currently under construction, will accommodate many budget flights, including those from IndiGo. Thus, like Kuala Lumpur International Airport, budget and full-service airlines will coalesce around a single facility.[45]

Low-cost-carriers have also proliferated in the Persian Gulf, though not at the same pace as in India, Southeast Asia, or along the Singapore-Australia corridor. John Bowen cites, among other factors, the preference among most Arab travelers

---

42 Initially Temasek, Singapore's governmental investment agency, partnered with Qantas on their Jetstar Asia venture based in Singapore. See *Financial Times*, "Qantas Takes a Triple Jump Beyond the Lucrative 'Kangaroo Route," May 31, 2004, Factiva.

43 Debra Schifrin, Glenn Carroll and Jesper Sorensen produced an important case study of SIA's budget carrier in "Scoot: Singapore Airlines' Low-Cost Carrier," Stanford Graduate School of Business, February 28, 2020, https://med.stanford.edu/content/dam/sm/CME/documents/SM321-20-1-.pdf, accessed January 9, 2023.

44 For a thorough discussion of IndiGo and the Indian aviation market in the recent past see Vishwajeet, *The IndiGo Story*.

45 NDTV.com, "Delhi Airport's Terminal-1 Cannot Be Exclusively Given to IndiGo: High Court," February 13, 2013, https://www.ndtv.com/india-news/terminal-1-cannot-be-exclusively-given-to-indigo-delhi-high-court-1812326, accessed January 9, 2023.

for luxury service, a quality in shorter supply on low-cost carriers (see chapter one regarding the preferences of Arab tourists).[46] The story of low-cost-carriers and their impact on hubbing in the United Arab Emirates reprises the controversial story of Dubai's entrance to high-stakes civil aviation, despite the presence of a viable airport in Sharjah. Affronted by Dubai's insistence that the new facility operate on "open skies model," Sharjah's leaders liquidated its flights on Gulf Aviation and shuddered its aging airfield.

Twenty years later, in 1977, however, Sharjah inaugurated a new domed airport that entertained the occasional commercial flight, including the Concorde from time to time.[47] It would be some decades later that the Sharjah Sheikh founded Air Arabia, which would be based at the new airport. Although it would not compete with a rising Emirates for long-haul preeminence, it did cater to students, tourists, and, most importantly, Southeast Asian and India workers that toiled in nearby Dubai.[48] Nevertheless, Sharjah's airport was only a secondary terminal in the sense that before Dubai inaugurated its own low-cost-carrier, flydubai, cost conscious travelers made the short bus ride to Sharjah. Ultimately, Dubai International Airport's status as a gateway to the Persian Gulf was never challenged, excepting perhaps in the percentage of *regional* market share captured by Air Arabia. However, neither of Dubai's current airports will be considered "secondary." Finally, the tiny thirteen room hotel operated by Air Arabia in the Sharjah International Airport poses little threat to the phalanx of hotels in nearby Dubai.

Conclusion

In summary, the advent of low-cost-carriers in Asia accelerated the penetration of commercial aviation in the Indo-Pacific. Most importantly, AirAsia Malaysia, Lion Air, Cebu Pacific, IndiGo Airlines and the like, seized the initiative for budget travel from national flag carriers. The low-cost-carriers and premium budget carriers (like Scoot Airlines, AsiaAir X, and Jetstar), which operate two-class configuration aircraft on medium and long haul, will also challenge the leadership of

---

46 Bowen, *Low Cost Carriers in Emerging Countries*, 146.

47 *Airports International,* "UAE Airports: As a Result of the UAE Becoming an Increasingly Successful Player in the Tourism and Commercial Markets, Its Airports are Working on some of the World's Greatest Development Plans," March 1, 2004, Factiva.

48 Katherine Zoepf, "No-Frills Air Arabia Creates a New Passion for Travel in the Persian Gulf," *The New York Times*, June 1, 2006, https://www.nytimes.com/2006/06/01/travel/01webletter.html, accessed January 9, 2023.

legacy carriers over longer routes. However, as Jayne Hrdlicka, then CEO of Jetstar noted in 2016, "If it's a 15-hour flight or a 13-hour flight, well then, you'll probably want a full-service experience. You're probably more prone to pay a bit more for that. But when you're looking anything from a five to 10-hour flight, I think [a premium budget airline like Jetstar] is a good experience."[49]

In terms of the hubbing model, Asian, Gulf, and Australasian airport authorities consolidated LCCs and legacy carriers in central hubs (such as at Dubai, Singapore, and Kuala Lumpur), This model was based on the paucity of secondary airports and a proclivity to maximise revenues from an acquisitive middle class. This stands in contrast to the pioneering of the secondary airport strategy tested in Europe and the United States. Nevertheless, as Max Hirsh and Richard Nieuvelle both argue, those same authorities must be sensitive to infrastructure strategies that consider the needs of working and middle-class passengers.

---

49 Sumers.

Conclusion

In March 2020, the footfall of millions of tourists, their voices, and the combined creaking of their carry-on roller bags, fell silent in aviation hubs across the Indo-Pacific with the spread of the COVID pandemic. Hubs in the Indo-Pacific were unique in that dueling airports bankrolled new ways to draw transit passengers from one opulent airport to their own. The costs had been staggering; over a billion dollars in the case of Singapore Changi Airport's Jewel. Passengers who once climbed through an HSBC sponsored forest, foraging for *haute couture* and *haute cuisine* in the halls outside the interior Eden, vanished in quick succession as nations sealed themselves off from incoming flights.

The onset of COVID provided an apt point of closure to a phase of travel—the hub-based transit experience—whose purported death sentence was read even before the pandemic. Conventional wisdom held that with advances in aviation technology, point-to-point air travel would render obsolete the hub what with its layovers and lethargy. More precisely, Qantas Airlines green-lit Project Sunrise, later approved by its board in 2022 for inauguration in 2025, which would offer non-stop service from Sydney, Australia, to London, England, on a twenty-hour flight.

Whatever the appeal (or lack thereof) of such efficiencies, the legacies of hubbing remain with us. It is likely that many travelers will not want to sit in an aluminum or composite-materials fuselage for twenty hours. Additionally, hubbing seems the only viable method for frequent journeys to parts of the world that do not merit high frequency point-to-point traffic (particularly in Africa, Latin America, and some secondary cities in Asia, for example).

This book historicises the hub experience, as well as its relationship to hometown, long-haul airlines, and iconic hotels. The first chapter examined what was one of the initial efforts to conceptualise the hubbing phenomenon, in that case by the Frank Lloyd Wright Foundation at the invitation of the Shah of Iran on the island of Kish in the late 1960s. While others had envisioned aviation hubs, the Taliesin West architects conducted site visits and laid out a nexus between intercontinental flight, upscale accommodations, and almost inconceivably outrageous attractions in the heart of the Persian Gulf. While the Shah considerably scaled back development at Kish given the costs, the blueprint resonated independently at other places in the Gulf, including Dubai, Doha, Bahrain, and Abu Dhabi.

At the same time, city-state leaders and their hand-picked consultants articulated a unique strategy for hosting air passengers near the mouth of the Malacca Straits. Situated on a central path from Europe to Southeast Asia, Singapore emerged as more than a military redoubt in the mid-twentieth century. Prior to

https://doi.org/10.1515/9783111326641-007

Table 1: The World's Best Airports, 2010–2019 (Rankings).[1]

Airport/	2019	2018	2017	2016	2015	2014	2013	2012	2011	2010
Singapore	1	1	1	1	1	1	1	2	2	1
Haneda	2	3	2	4	5	6	9	14	17	–
Seoul Incheon	3	2	3	2	2	2	2	1	3	2
Doha	4	5	6	10	–	–	–	–	–	–
Hong Kong	5	4	5	5	4	4	4	3	1	3
Nagoya	6	7	7	6	7	12	13	10	11	13
Munich	7	6	4	3	3	3	6	6	5	4
Heathrow	8	8	9	8	8	10	10	11	16	21
Tokyo Narita	9	11	14	11	14	16	17	17	19	17
Zurich	10	9	8	7	6	8	7	7	7	6

that point it was the key transit point linking commercial aviation between India and Australia along what would be known as the Kangaroo Route—running between London and its antipodal outpost, Sydney.

When autonomy, then independence, came to Singapore, its early leaders from the People's Action Party, including Goh Keng Swee, eschewed international aid agencies as the cornerstone of a viable tourism development programme, electing instead to focus on transit passengers who might make Singapore the hub of their own Southeast Asian travels. Subsequently, as relations between Singapore and Malaysia deteriorated in the early 1970s, Singapore Airlines acquired several long-range Boeing 707s and focused almost exclusively on long-haul travel in and between Asia, Europe, and the United States.

By the mid-1970s, exponential growth in passenger movements through Singapore signaled the need for a new airport at the old Changi airfield. The airport evolved into a customer-centric transit point renowned throughout the world for its high quality of service, growing shopping options, and passenger amenities that made the transit experience more pleasant. These efforts culminated with

---

1 Sources: Skytrax, https://www.worldairportawards.com/worlds-best-airports-announced-2019/, https://www.worldairportaward.com/worlds-top-100-airports-2018/, https://www.worldair portawards.com/best-airports-of-2017-unveiled-at-world-airport-awards/, https://www.worldair portawards.com/the-worlds-top-100-airports-2015/, https://www.worldairportawards.com/the-worlds-top-100-airports-2012/, https://www.moodiedavittreport.com/skytrax-names-changi-air port-as-worlds-best-airport-2010-ahead-of-incheon-international/, all accessed November 2, 2021. No U.S. airport has appeared in the top three of Skytrax lists of the world's best airports since records started in 1999. For a broader array of Skytrax data, see https://en.wikipedia.org/wiki/Skytrax, accessed November 2, 2021.

the construction of the Jewel Project in 2019, a testament to the continued power of airports to invoke wonder.

As chapter two also illustrates, Changi faced stiff competition from the traditional air hub of East Asia, Hong Kong, sitting at the fulcrum of the Asia Pacific and Southeast Asia. Singapore enjoyed the advantage of political independence over Hong Kong, whose status as a colony, much less air hub, remained in limbo during its handover from the British Crown to China in 1997. Now Hong Kong faces as much scrutiny for its perceived willingness to cooperate (or compete) with surrounding Pearl River Delta airports as it does for its political relationship with the People's Republic of China in the run-up to 2047.[2]

Table 2: The World's Top Airlines, 2009–2019.[3]

Airline/Hub	2019	2018	2017	2016	2015	2014	2013	2012	2011	2010	2009
Qatar (Doha)	1	2	1	2	1	2	3	1	1	3	4
Singapore (Changi)	2	1	2	3	2	3	3	3	2	2	2
ANA (Narita)	3	3	3	5	7	6	4	5	11	–	–
Cathay (HK)	4	6	5	4	3	1	6	4	4	4	1
Emirates (Dubai)	5	4	4	1	5	4	1	8	10	8	5
EVA Air (Taipei)	6	5	6	8	9	12	12	13	16	–	–
Hainan (Beijing)	7	8	9	12	22	19	19	20	23	–	–
Qantas (Sydney)	8	11	15	9	10	11	10	15	8	7	6
Lufthansa (Munich)	9	7	7	10	12	10	11	14	15	–	–
Thai (Bangkok)	10	10	11	13	19	14	15	9	5	9	10

2 The relationship between Chek Lap Tok and surrounding airports at Guangzhou, Shenzhen, Zhuhai, and Macau, has been a persistent question since even before the new Hong Kong airport opened in 1998. See, for example, Peter Lim's article for the *Agence France-Presse*, "Hong Kong's New Airport Opens into Uncertain Future," July 5, 1998, Factiva. The Asian economic crisis was cited as one deterrent to new tourism flows through the city-colony. At the same time, Lim noted, "China is also going ahead with the opening of new airports including Shanghai next year and Guangzhou in 2001." Guangzhou shot up the charts of mainland China's busiest international, as well as domestic airports, by the 2020s. It remains to be seen how well the airports will coordinate their futures, particularly given the sometimes-zero-sum approach of competition between rival mainland airports.

3 Source: Skytrax, https://www.worldairlineawards.com/worlds-top-100-airlines-2019/, https://www.worldairlineawards.com/world-airline-awards-announced-2017/, https://www.worldairlineawards.com/the-worlds-top-100-airlines-2016/, https://www.worldairlineawards.com/the-worlds-top-100-airlines-2014/, https://www.worldairlineawards.com/the-worlds-top-100-airlines-2012/, https://www.rankingthebrands.com/The-Brand-Rankings.aspx?rankingID=270&year=506.

To the west, Dubai International Airport grew in tandem with Changi, pioneering the concept of massive airport-centric duty-free shopping, an idea adapted from the Shannon, Ireland, airport. With its own hometown airline, Emirates Airlines, Dubai competed with Singapore and Hong Kong for Asian passengers headed to Europe, but also established African and Middle Eastern networks that proved complimentary to those of its rival hubs to the East. Shanghai, Mumbai, Bangkok, and Kuala Lumpur also endeavored to establish their own entrepots for tourists, with varying levels of success.

Table 3: Top 10 Global Airports by Retail Sales, 2019.[4]

Ranking	Airport	Annual Traffic (passengers)	Retail Sales(dollars)
1	Seoul Incheon	70 million	2.43 billion
2	Singapore Changi	68.3 million	2.06 billion
3	Dubai International	86.5 million	2.03 billion
4	Shanghai Pudong	76 million	1.23 billion
5	Beijing Capital	80.9 million	1.07 billion
6	London Heathrow	80.8 million	946 million
7	Paris Charles de Gaulle	76.2 million	860 million
8	Hong Kong International	60.9 million	759 million
9	Tokyo, Narita	44.3 million	603 million
10	Istanbul Ataturk	52.6 million	560 million (based on 9 months in new airport)

As chapter three detailed, hotel development throughout Asia fed off the rise of commercial aviation. Bypassing traditional ocean-going stops along the Suez to Hong Kong route, including Mumbai and Colombo, fledgling airlines patronised luxury hotels along their flight path or built modern hotels for their passengers, particularly in the Persian Gulf. With decolonisation, Asian hotel developers, principally hailing from the wheelhouse of commodities trading, built their own hospitality networks to meet the throngs of tourists brought by jumbo jets. The style of these hotels often reflected Asian themes and fulfilled largely commercial purposes, in contrast to the more geopolitically oriented American hotel chains that lightly dotted the East Asian Pacific. Finally, hubs became destinations. A

---

4 *TR Business: the leading provider of duty free and world retail news,* August 2020 https://content.yudu.com/web/42i3x/0A42jbv/Aug2020/html/index.html?refUrl=https%253A%252F%252Fcontent.yudu.com%252Fweb%252F42i3x%252F0A42jbv%252FAug2020%252Findex.html&page=01.

hometown airline, a burgeoning airport, and opulent casinos, seven-star service, and technological showpieces drew a global clientele to the Indo-Pacific.

Chapter four explored how the Kangaroo Route thrived on the traffic of Australians moving between their remote continent and the British metropole. These tourists, airline staff, and businesspeople, brought their collective experience to bear back home. Intercontinental flight emerged as the standard against which Australian airlines and airports were judged. Innovation often followed on this longest of feedback loops. One recent example might be Kingsford-Smith's modest luxury "precinct." The arcade includes thirteen high-street brands including Dior, Bottega Venata, Balenciaga, Prada and Louis Vuitton's "first travel retail store in the Southern Hemisphere."[5] Ultimately, Australia's integration into global aviation would be both conventionally Western, with its tight connections with British Overseas Airways Corporation, as well as innovatively Asian, with strategies in Japan and Thailand during the 1980s, which opened cultural horizons to travelers, politicians, and businesspeople alike.

More recently, low-cost-carriers in Southeast Asia and the broader Indo-Pacific have forced a paradigm-shift upon the commercial aviation world, and by extension, accelerated the hubbing process. Tony Fernandes's Air Asia conglomeration of subsidiaries led the charge in Southeast Asia and created new opportunities for the working- and middle-class to travel short, medium, or long distances. Chapter five contends that while low-cost-carriers followed a core set of practices that delivered lower prices per kilometer to passengers, its impact on aviation infrastructure in Asia was unique. Airports were more than simply a reflection of airline owner demands. In places like Kuala Lumpur, Singapore, and Hong Kong, LCC airlines operated out of traditional terminals that balanced increasingly automated services for airlines along with the high-end shopping that many passengers in a brand-sensitive society expected. In effect, while LCCs have disrupted the domestic primacy of full-service legacy carriers, they have reinforced the hubbing model by virtue of their presence at traditional airports. What's more, LCCs have found traditional hubs ideal locations to attract new passengers. When added to the reality that legacy carriers flying widebody jets better met the needs of intercontinental travelers, it is doubtful that hubbing will disappear in Asia. Most likely, the two will exist side-by-side in increasingly state of the art airports that are also designed with the flexibility to adapt the passenger needs with innovations in aviation.

---

5 Christopher Kelly, "Gallery: Inside Louis Vuitton's Sydney Airport Store," *Ragtrader*, November 15, 2022, Factiva. The store's furnishing was decided "Eastern" in taste, featuring "an architectural concept first implemented by Peter Marino in Ginza Namiki Dori, Tokyo, Japan."

Finally, the duties of the historian do not include business forecasts, but, while hubbing came to something of a pause during the COVID pandemic, the continued rise of African and West Asian destinations—which may not justify daily point-to-point widebody service—makes financial sense to airlines deploying the hubbing model. The historical staying power of Cathay Pacific, Singapore Airlines, Emirates and Qatar Airways, not to mention the durability of their cities of origin, will likely perpetuate some variation on the hubbing model.

From a broader historical perspective, the story of "hubbing for tourists" affirms the narrative of a creative and economically powerful Asia and Indo-Pacific reclaiming their place on the world stage. It is fitting that one of the most influential histories of the past twenty-five years, Kenneth Pomeranz's *The Great Divergence: China, Europe, and the Making of the Modern World Economy* (Princeton University Press, 2000) examines shifts in economic power between West and East. At that moment, somewhere during the eighteenth century, European living standards eclipsed those of China and Asia more broadly, hence the designation of a "great divergence." The story we read here, though on a very narrow basis, suggests a growing *convergence* between the Indo-Pacific, Europe, and the United States even as we speak. Fittingly, this is about more than simply money and power. It is about creativity and enduring institutions that stretch across time and human experience. If the port, the *caravanserai* (an Ottoman or Mughal inn for overland travellers), and the bazaar (or marketplace) epitomised the mobility of people and wealth prior to the Great Divergence, then it is the reconfiguration of those symbols into new institutions that sets the pace for the rest of the twenty-first century. It's no surprise that these hybrid entities flourish in Asia.

Bibliography

Archives and Libraries

National Archives of Singapore, Singapore.
Special Collections, School of Oriental and African Studies, University of London, London, United Kingdom.
British National Archives, Kew, United Kingdom.
Australian National Archives, Canberra, Australia.
Australian National Archives, Melbourne, Australia.
Special Collections, National Library of Australia, Canberra, Australia.
Library of Congress, Washington D.C.
Taliesin Associated Architects Archives, Frank Lloyd Wright Foundation, Scottsdale, Arizona.
Asian Bank of Development, Manila, Philippines.
Hong Kong Heritage Project, Kowloon, Hong Kong, S.A.R., China.
Hong Kong Government Records Office, Kowloon, Hong Kong, S.A.R., China.
Swire and Sons Corporate Archives, Cambridge House, Hong Kong, S.A.R., China.
New Zealand National Archives, Wellington, New Zealand.

Primary Source Databases

Arabian Gulf Digital Archives, https://www.agda.ae/en. All documents accessed from this archive originate at the British National Archives, but have been scanned by authorities in the Gulf.
Trove.com. National Library of Australia. Australian newspaper database. Note: All Australian newspapers have been accessed through Trove *unless otherwise noted*.
Factiva. Global newspaper database.Pro-quest Newspaper Database. Global newspaper database.
World Bank Documents.

Other Sources

Adalet, Begüm Adalet. *Hotels and Highways: The Construction of Modernization Theory in Cold War Turkey*. Palo Alto: Stanford University Press, 2018.
"Air France Far East Timetable U.K. edition." March 30, 1938. In: Airline Timetable Images. Accessed June 24, 2022. http://timetableimages.com/ttimages/af3803fe.htm.
Allchin, F.R. *Cultural Tourism in India: Its Scope and Development with Special Reference to the Monumental Heritage*. Paris: UNESCO, 1969. Accessed June 29, 2022. https://unesdoc.unesco.org/ark:/48223/pf0000008820?posInSet=1&queryId=4c631774-ee2d-4e96-9bc8-39accf73e07e.
Arch Daily. "The World's First Quarry Hotel Opens in China, Designed by Jade + QA." July 2, 2022. https://www.archdaily.com/910409/the-worlds-first-quarry-hotel-opens-in-china-designed-by-jade-plus-qa.
Archis, AMO, and Moutamarat. *Al Manakh*. Amsterdam: Archis, 2007.

https://doi.org/10.1515/9783111326641-008

Aslanian, Sebouh David. *From the Indian Ocean to the Mediterranean: The Global Trade Networks of Armenian Merchants from New Julfa*. Berkeley: University of California Press, 2014.

Augustin, Andreas. *The Imperial: New Delhi*. London: The Most Famous Hotels in the World, Ltd.

Augustin, Andreas. *The Incredible Tale of the Legendary Strand, The Most Famous Hotel of Rangoon*. London: The Most Famous Hotels in the World, Ltd., 2017.

Augustin, Andreas, and Andrew Williamson. *The Oriental Bangkok*. The Most Famous Hotels in the World. London: The Most Famous Hotels in the World, Ltd., 1996.

B1M. "Shanghai's Underwater Quarry Hotel." Accessed July 2, 2022.https://www.youtube.com/watch?v=DkTDjx2jLj8.

Balint, Ruth. "Epilogue: The Yellow Sea." In: David Walker, ed. *Australia's Asia: From Yellow Peril to Asian Century*. Perth: UWA Publishing, 2012.

Barr, Michael D. *Singapore: A Modern History*. London: Bloomsbury, 2019.

Binggeli, Urs and Mathieu Weber. "A Short Life in Long Haul for Low-Cost Carriers." Travel, Trasport & Logistics. McKinsey&Company. Accessed January 9, 2023. https://www.mckinsey.com/~/media/McKinsey/Industries/Travel%20Transport%20and%20Logistics/Our%20Insights/A%20short%20life%20in%20long%20haul%20for%20low%20cost%20carriers/A_short_life_in_long_haul_for_low%20cost_carriers.pdf.

Bergdoll, Berry Bergdoll and Jennifer Gray, eds. *Frank Lloyd Wright: Unpacking the Archive*. New York: Museum of Modern Art, 2017.

Betta, Chiara. "Orientals to Imagined Britons: Baghdadi Jews in Shanghai." *Modern Asian Studies* 37 (2003): 999–1023.

Bickers, Robert. *China Bound: John Swire & Sons and Its World, 1816–1980*. London: Bloomsbury, 2020.

Bill, James A. *The Eagle and the Lion: The Tragedy of American-Iranian Relations*. New Haven, CT: Yale University Press, 1988.

Blainey, Geoffrey. *A Shorter History of Australia*. Revised and updated. North Sydney: Vintage, 2009.

Blainey, Geoffrey. *The Tyranny of Distance: How Distance Shaped Australia's History*. 2001.

Box, Rachel. "Airport on the Move? The Policy Mobilities of Singapore Changi Airport at Home and Abroad." *Urban Studies* 52 (2015).

Bollen, Jonathan. *Touring Variety in the Asia Pacific Region, 1946–1975*. Transnational Theatre Histories. Palgrave Macmillan, 2020.

Bose, Sugata. *A Hundred Horizons: The Indian Ocean in the Age of Global Empire*. Cambridge, Massachusetts: Harvard University Press, 2006.

Bowen, John. *Low-Cost Carriers in Emerging Countries*. Amsterdam: Elsevier, 2019.

Brook, Daniel. *A History of Future Cities*. New York City: W.W. Norton, 2013.

Center for Asia Pacific Aviation. "Spectrum Report." Accessed June 8, 2022. https://www.airvistara.com/content/dam/airvistara/global/english/common/documents/Spectrum_Report.pdf.

Chang, T.C. "Regionalism and Tourism: Exploring Integral Links in Singapore." *Asia Pacific Viewpoint* 39 (1998): 73–94.

Chang, T.C. "Sustaining Singapore Tourism Through the Years: Policy, People, Place." Lawal Mohammed Marafa and Chung-Shing Chan, eds. 188–199. *Sustainable Tourism in Asia: People and Places* (Newcastle on Tyne, UK: Cambridge Scholars Publishing, 2019.

Checchi and Company. *A Proposal to Assist Alaska to Develop its Tourism Potentials*.

Clement, Harry G. *The Future of Tourism in the Pacific and Far East*. Washington D.C.: Department of Commerce, 1961.

Cornell Hotel and Hospitality Management. "Travel in the 1970s." 10 (1970).

Cornell Hotel and Restaurant Administration Quarterly, Treasure Trove 1964. Number 4 (1964): 30–41.

Corporate Travel Community. "AirAsia's Tony Fernandes Again Laments the Lack of Low-Cost Terminals in Malaysia; Points to Japan for Inspiration." Accessed January 17, 2023. https://corporatetravelcommunity.com/analysis/airasias-tony-fernandes-again-laments-the-lack-of-low-cost-terminals-in-malaysia-points-to-japan-for-585492.

Crotty, David. *Qantas the Empire Flying Boat*. Stamford Lincolnshire, UK: Key Publishing, 2021.

Davidson, Christopher. *Dubai: The Vulnerability of Success*. New York City: Oxford University Press, 2008.

Day, Kirsten. "The Shanghai Paradox." The Mediated City Conference. Woodbury University. 1–3 April 2014. Accessed August 28, 2018. http://architecturemps.com/wp-content/uploads/2013/09/mc_conference_day_kirsten1.pdf.

De Neufville, Richard. "Low-Cost Airports for Low-Cost Airlines: Flexible Design to Manage the Risks." *Transportation Planning and Technology* 31.

Department of the Prime Minister and Cabinet. *Australia in the Asian Century*. White Paper Canberra: Department of the Prime Minister and Cabinet, October 2012. Accessed September 3, 2022. https://apo.org.au/node/31647.

Doshi, Tila andPeter Coclanis. "The Economic Architect: Goh Keng Swee," In: Kevin Yi Tan and Lam Peng Er, eds. *Lee's Lieutenants: Singapore's Old Guard*. Revised edition. Singapore: Straits Times Press, 2018. 80–109.

Duara, Prasenjit Duara. "Asia Redux: Conceptualizing a Region for our Times." *The Journal of Asian Studies* 69 (2010).

Elsheshtawy, Yassar. *Dubai: Behind an Urban Spectacle*. London: Routledge, 2013.

Fallows, James. *China Airborne: The Test of China's Future*. New York City: Vintage, 2012.

Fernandes, Tony. *Flying High: My Story from AirAsia to QPR*. London: Portfolio Penguin, 2017.

Francis, Graham, Alessandro Fidato, and Ian Humphreys. "Airport-Airline Interaction: The Impact of Low-Cost Carriers on Two European Airports." *Journal of Air Transport Management* 9 (2003).

Fregonese, Sara and Adan Ramadan. "Hotel Geopolitics: A Research Agenda" *Geopolitics* 20 (2015): 793–813.

Furlough, Ellen andRosemary Wakeman. "La Grande Motte: Regional Development, Tourism, and the State" In: Baranowski, Shelley Osmun, and Ellen Furlough, eds. *Being Elsewhere: Tourism, Consumer Culture, and Identity in Modern Europe and North America*. Ann Arbor: University of Michigan Press, 2001. 348–372.

Ghattas, Kim. "How Lebanon Transformed Anthony Bourdain." *The Atlantic*. June 9, 2018. Accessed November 5, 2018. https://www.theatlantic.com/international/archive/2018/06/how-lebanon-transformed-anthony-bourdain/562484/.

Goh Keng Swee. "Speech by Dr. Goh Keng Swee, Minister for Finance at the Annual Dinner of the Singapore Hotels and Restaurants Association at the Golden Lotus, Hotel Malaysia, on Friday, 24th October 1969 at 8:00pm." Accessed June 16, 2021. https://www.nas.gov.sg/archivesonline/data/pdfdoc/PressR19691024d.pdf.

Grescoe, Taras. *Shanghai Grand: Forbidden Love and International Intrigue in a Doomed World*. New York: St. Martin's Press, 2016.

Gross, Sven and Michael Lück, eds. *The Low Cost Carrier Worldwide*. Abingdon Oxon, UK, 2013.

Gunn, John. *The Defeat of Distance, Qantas, 1919–1939*. St. Lucia, Queensland: Queensland University Press, 1985.

Gunn, John. *High Corridors: Qantas, 1954–1970* St. Lucia, Queensland: Queensland University Press, 1988.

Gupte, Pranay. *Dubai: The Making of a Megalopolis*. New York: Viking, 2011.

Hakar, Ajit N. *Bite the Bullet: Thirty-Four Years with ITC*. Viking, 1993.

Harris, Kerr, Forster & Company and Stanton Robbins & Company, Incorporated. *Australia's Travel and Tourist Industry*, New York, 1965.

Haryopratomo, Aldi, Sanja Kos, Lavin Samtani, Sheela Subramanian, and Jay Verjee. "The Dubai Tourism Cluster: From the Desert to the Dream." May 6, 2011. Accessed January 24, 2022. https://www.isc.hbs.edu/Documents/resources/courses/moc-course-at-harvard/pdf/student-projects/UAE_(Dubai)_Tourism_2011.pdf.

Hazbun, Waleed. *Beaches, Ruins, Resorts: The Politics of Tourism in the Arab World*. Minneapolis: University of Minnesota Press, 2008.

Henderson, Joan C. "Case Study: Uniquely Singapore? A Case Study in Destination Branding." *Journal of Vacation Marketing* 13 (2007): 261–274.

Henderson, Joan C. "Conserving Colonial Heritage: Raffles Hotel in Singapore," *International Journal of Heritage Studies*, 7(2001).

Henderson, Joan C. "Destination Development." *Journal of Travel & Tourism Marketing* 20: 33–45.

Henderson, Joan C. "Tourism and Development in Singapore." In: Eduardo Fayos-Sola, ed., *Tourism as an Instrument for Development: A Theoretical and Practical Study*. Emerald Books, 2014. 84–88.

Henderson, Joan C. "Destination Development and Transformation: 50 years of Tourism After Independence in Singapore." In: *International Journal of Tourism Cities*, 1: 269–281.

Heseltine, Michael Heseltine. *Life in the Jungle: My Autobiography*. Hodder and Stoughton Ltd, 2000.

Hickson, Ken. *Mr. SIA: Fly Past*. Singapore: World Scientific, 2015.

Hirsh, Max. *Airport Urbanism: Infrastructure and Mobility in Asia*. Minneapolis: University of Minnesota Press, 2016.

Ho, Engseng. "Inter-Asian Concepts for Mobile Societies." *The Journal of Asian Studies*. 76 (2017), 907–928.

Hong Kong and Shanghai Hotels, Limited. *Beyond Hospitality*. Hong Kong: Marshall Cavendish, 2010.

Hong Kong Tourism Association. *Report of the Board of Management of the Hong Kong Tourism Association*, 1959/60.

Howe, Yoon Chong. "Speech by the Minister of Defence, Mr. Howe Yoon Chong, at the Opening of the Singapore Changi Airport on Tuesday, December 29, 1981 at 3:30pm." Accessed June 16, 2021. https://www.nas.gov.sg/archivesonline/data/pdfdoc/hyc19811229s.pdf.

Isenstadt, Sandy andKishwar Rivzi. "Introduction: Modern Architecture and the Middle East: The Burden of Representation." In: *Modernism and the Middle East: Architecture and Politics in the Twentieth-Century*, Sandy Isenstadt and Kishwar Rivzi, eds. Seattle: University of Washington Press, 2008.

James, Kevin J., A.K. Sandoval Strausz, Daniel Maudlin, Maurizio Peleggi, Cedric Humair, and Molly W. Berger. "The Hotel in History: Evolving Perspectives." *Journal of Tourism History* 9 (2017).

John Desmond Ltd. "The Shangri-La at the Shard." Accessed June 13, 2022. https://www.johndesmond.com/blog/design/shangri-la-shard/.

Karkaria, Bachri J. *Dare to Dream: The Life of M.S. Oberoi*. New Delhi: Penguin Books, 2006.

Kasarda, John D. and George Lindsay. *Aerotropolis: How We'll Live Next*. New York City: Farrar, Straus, and Giroux, 2011.

Kaufman, Jonathan. *The Last Kings of Shanghai: The Rival Jewish Dynasties that Helped Create Modern China*. New York: Penguin, 2020.

Keshavarzian, Arang. "Geopolitics and the Genealogy of Free Trade Zones in the Persian Gulf." *Geopolitics* 15 (2010), 263–289.

Kharga, Namrata Kharga. "Mount Everest Hotel Darjeeling Echoes of the Forgotten Glory." Accessed June 29, 2022. https://ramblingcraft.com/mount-everest-hotel-darjeeling-echoes-of-the-forgotten-glory-2/.

KKS Group. "Works." Accessed June 10, 2022. https://www.kkstokyo.co.jp/en-works/archive/?id=1960#a1960.

KKS Group. "Philosophy." Accessed June 10, 2022. https://www.kkstokyo.co.jp/en/philosophy/.

KKS Group. "Service." Accessed June 10, 2022. https://www.kkstokyo.co.jp/en/service/.

Klein, Christina. *Cold War Orientalism: Asia in the Middlebrow Imagination, 1945–1961*. Berkeley: University of California Press, 2003.

Koh, Tommie. *Singapore the Encyclopedia*. Singapore: Editions Didier Millet and the National Heritage Board, 2006.

Krane, Jim. *City of Gold: Dubai and the Dream of Capitalism*. New York City: St. Martin's Press, 2009.

Kwa, Chong Guan, Derek Heng, Peter Breschberg and Tan Tai Yong. *Seven Hundred Years: A History of Singapore*. Singapore: National Library Board, 2019.

Kuok, Robert Hock Nien and Andrew Tanzer. *A Memoir*. Singapore: Landmark Books, 2017.

Lala, R.M. *Beyond the Last Blue Mountain: The Life of J.D.R. Tata*. Penguin, 2017.

Leary, William M. Jr. *The Dragon's Wings: The China National Aviation Corporation and the Development of Commercial Aviation in China*. Athens: University of Georgia Press, 1976.

Lee Hsien Loong. "Prime Minister Lee Hsein Loong's National Day Rally 2013 (English)," August 18, 2013. Accessed June 15, 2021. https://www.pmo.gov.sg/Newsroom/prime-minister-lee-hsien-loongs-national-day-rally-2013-english.

Lee Kuan Yew. "Address by the Prime Minister, Mr. Lee Kuan Yew, at the SIA's 30th Anniversary Dinner, Held at the Neptune Theatre Restaurant, on Sunday, 1st May, 1977." Accessed June 14, 2021. https://www.nas.gov.sg/archivesonline/data/pdfdoc/lky19770501.pdf.

Lee Kuan Yew. "Speech by Mr. Lee Kuan Yew, Minister Mentor, at Launch of the New CAAS and the Changi Airport Group, July 1, 2009, 4:00pm at Changi Airport Terminal 3." Accessed June 15, 2021. https://www.pmo.gov.sg/Newsroom/speech-mr-lee-kuan-yew-minister-mentor-launch-new-caas-and-changi-airport-group-01-july.

Lee Kuan Yew. "Transcript of an Interview with Singapore's Prime Minister, Mr. Kee Kuan Yew, by NZBC's Hylda Bamber, Recorded in Dunedin, New Zealand, on 13th March, 1965)." Accessed June 15, 2021. https://www.nas.gov.sg/archivesonline/data/pdfdoc/lky19650313.pdf.

Le Renard, Amelie. *Western Privilege: Work, Intimacy, and Post-Colonial Hierarchies in Dubai*. Jane Kuntz, trans. Palo Alto, California: Stanford University Press, 2021.

Lien Ying Chow. *From Chinese Villager to Singapore Tycoon; My Life Story*. With Louis Kraar. Entrepreneurs of Asia Series. Singapore: Times Books International,1992.

Lin Yong Nyuk. "Mr. Yong Nyuk Lin, Minister for Communications at the Welcoming Ceremony of the First 2 SIA-BOEING 747s at Paya Lebar Airport on Monday 3rd September 1973 @ 1600 Hours, "SIA Joins the Jumbo Jet League." Accessed June 16, 2021. https://www.nas.gov.sg/archiveson line/data/pdfdoc/PressR19730903.pdf.

Lin Yong Nyuk. "Statement of R. Yong Nyuk Lin, Minister for Communications, at the Press Conference Held at Paya Lebar Airport on Friday 4/12/70 at 1240 Hours, "'Paya Lebar International Airport is Being Geared up for the 1980s!'" Accessed June 16, 2021. https://www.nas.gov.sg/archivesonline/data/pdfdoc/PressR19701203a.pdf.

Liu, Gretchen. *Raffles Hotel*. Singapore: Landmark Books, 1992.

Lohmann, Guilherme Lohmann, Sascha Albers, Benjamin Koch, and Kathryn Pavlovich. "From hub to tourist destination – An explorative study of Singapore and Dubai's aviation-based transformation." *Journal of Air Transport*. Accessed January 24, 2022. https://doi.org/10.1016/j.jair traman.2008.07.004,https://research-repository.griffith.edu.au/bitstream/handle/10072/56170/89415_1.pdf?sequence=1.

Lück, Michael and Sven Gross. "Low Cost Carriers in Australia and New Zealand." In: Sven Gross and Michael Lück, eds. *The Low Cost Carrier Worldwide*. Abingdon, Oxon, UK: Routledge, 2014.

Mama, Hormuz P. *Second Airport: What Mumbai Must Learn from International Experience*. Mumbai: Observer Research Foundation Mumbai, 2010.

Martí-Ibáñez, Félix. *Journey Around Myself: Impressions and tales of Travels Around the World: Japan, Hong Kong, Macao, Bangkok, Angkor, Lebanon*. New York: Clarkson N. Potter, Inc., 1966.

Martin, Lou. *Wings over Persia*. Fifth Edition. Victoria, British Columbia: Trafford, 2005.

McClean, Ian W. *Why Australia Prospered: The Shifting Sources of Economic Growth*. Princeton, N.J.: Princeton University Press, 2012.

Miller, Rory. *Desert Kingdoms to Global Powers: The Rise of the Arab Gulf*. New Haven, Connecticut: Yale University Press, 2016.

Ministry of Aviation. Government of India. *Report of Working Group on Civil Aviation for Formulation of Twelfth Five Year Plan (2012–1017)*. Accessed June 8, 2022. https://www.civilaviation.gov.in/sites/default/files/moca_001320.pdf.

Ministry of Communications and Information *Singapore 1982*. Singapore: Ministry of Communications and Information, 1982.

Ministry of Communications and Information, *Singapore 1983*. Singapore: Ministry of Communications and Information, 1983.

Ministry of Communications and Information, *Singapore 1984*. Singapore: Ministry of Communications and Information, 1984.

Ministry of Communications and Information. *Singapore 1985*. Singapore: Ministry of Communications and Information, 1985.

Ministry of Communications and Information. *Singapore 1986*. Singapore: Ministry of Communications and Information, 1986.

Ministry of Communications and Information. *Singapore 1987*. Singapore: Ministry of Communications and Information, 1987.

Ministry of Communications and Information. *Singapore 1988*. Singapore: Ministry of Communications and Information, 1988.

My Goh Chok Tong. "Speech by the Prime Minister, My Goh Chok Tong, at the Official Opening of the Singapore Changi Airport, Terminal 2, on Saturday, June 1, 1991, at 11:00AM." Accessed June 16, 2021. https://www.nas.gov.sg/archivesonline/data/pdfdoc/gct19910601.pdf.

Naira, E. Shailaja. *Goh Keng Swee: the master sculptor*. Singapore: SNP Editions, 2008.

Narayanan, Chitra. *From Oberoi to Oyo: Behind the Scenes with the Movers and Shakers of India's Hotel Industry*. New Delhi: Penguin, 2020.

National Archive of Singapore. *The 2nd Decade: National Building in Progress, 1975–1985*. Singapore: NAS, 2010.

Navarro, Alfredo Mena, Fernando Almeida Garcia, and Rafael Cortes Macias. "Evolution of Singapore Tourist Policy (1965–2015)." *Cuadernos de Turismo* 41 (2018), 703–706.

O'Connor, Kevin andKurt Fuelhart. "The Asia Pacific Region and Australian Aviation." In: *Air Transport in the Asia Pacific*. David Timothy Duval, ed. Farnham, UK: Ashgate, 2014.

Ooi, Can Seng. "State-Civil Society Relations and Tourism: Singaporeanizing Tourists, Touristifying Singapore." *Sojourn* 20 (2005), 249–272.

Oxford Economics. "Explaining Dubai's Aviation Model." In: *A Report for Emirates and Dubai Airport*. 2011. Accessed January 24, 2022. https://www.oxfordeconomics.com/my-oxford/projects/128910.

Peleggi, Maurizio. "Consuming Colonial Nostalgia: The Monumentalisation of Historic Hotels in urban South-East Asia." *Asia Pacific Viewpoint*, 26 (2005).

Peleggi, Maurizio. "The Social and Material Life of Colonial Hotels: Comfort Zones as Contact Zones in British Colombo and Singapore, ca. 1870–1930." *Journal of Social History*. 46 (2012): 124–153.

Perry, John Curtis. *Singapore: Unlikely Power*. New York City: Oxford University Press, 2017.

Pike, Steven and Filareti Kotsi. "Stopover Destination Image—Perceptions of Dubai, United Arab Emirates, among French and Australian Travellers." *Journal of Travel and Tourism Marketing* 35 (2018).

Pillay, J.Y. "The Man Behind Singapore Airlines." In: *Speaking Truth to Power: Singapore's Pioneer Public Servants*, Loke Hoe Yeong, ed. *The Singapore Story by the History-makers*. volume 1 Singapore: World Scientific.

Pirie, Gordon. "Incidental Tourism: British Imperial Air Travel in the 1930s." *Journal of Tourism History*. 1 (2009), 49–66.

Potter, James E. Potter. *A Room with a World View: 50 Years of InterContinentalHotels and its People, 1946–1996*. London: Weidenfeld & Nicolson, 1996.

Rajaratnam, Sinnathamby. *Singapore: Global City* (Singapore: Government Printing Office, n.d.), "Text of address by Mr. [Sinnathamby] Rajaratnam, Minister for Foreign Affairs to the Singapore Press Club on February 6, 1972," Accessed June 16, 2021. https://www.nas.gov.sg/archivesonline/data/pdfdoc/PressR19720206a.pdf.

Rehman, Habib with Sourish Bhattacharyya. *Borders to Boardroom: A Memoir*. New Delhi: Roli Books, 2014.

Reisz, Todd, ed. *Al Manakh: Continued*. Amsterdam: Archis, 2010.

Reisz, Todd. *Showpiece City: How Architecture Made Dubai*. Palo Alto, California: Stanford University Press, 2020.

Reza Pahlavi, Mohammad. *The White Revolution*. Second edition. Teheran: Imperial Pahlavi Library, 1967.

Robb Report. "How Flight Delays are Turning Airports into Luxury Malls." March 27, 2019. Accessed November 19, 2022. https://robbreport.com/lifestyle/news/luxury-retailers-airport-stores -2845150/.

Safdie, Moshe. "The Design Concept." In: *Jewel Changi Airport*. Sam Lubell, ed. Melbourne: The Images Publishing Group, 2020. 26–36.

Safdie, Moshe. "Rethinking the Public Realm." In: *Reaching for the Sky: The Marina Bay Sands Singapore*. ORO Editions, 2013.

Salazar, Noel B. "Tourism Imaginaries: A Conceptual Approach." *Annals of Tourism Research* 20 (2011).

Salem, Elie Adib. *Modernization without Revolution: Lebanon's Experience*. Bloomington: Indiana University Press, 1973.

Sarkissian, Margaret. "Armenians in South-East Asia." *Crossroads: An Inter-disciplinary Journal of Southeast Asian Studies* 3 (1987): 1–33.

Sarmadi, Bahzad. "'This Place Should Have Been Iran': Iranian Imaginings in/of Dubai." May 20, 2013. *Ajam Media Collective*. Accessed August 28, 2018. https://ajammc.com/2013/05/20/this-place-that-should-have-been-iran-iranian-imaginings-inof-dubai/.

Seng, Eunice. "Temporary Domesticities: The Southeast Asian hotel as (Re)presentation of Modernity, 1968–1973. *The Journal of Architecture*. 22 (2017), 1092–1136.

Seng, Loh Kah. "The British Military Withdrawal from Singapore and the Anatomy of a Catalyst." In: Derek Heng and Syed Muhd Khairudin Aljunied, eds. *Singapore in Global History*. Amsterdam: Amsterdam University Press, 2011.

Schifrin, Debra, Glenn Carroll and Jesper Sorensen. "Scoot: Singapore Airlines' Low-Cost Carrier" Stanford Graduate School of Business. February 28, 2020. Accessed January 9, 2023. https://med.stanford.edu/content/dam/sm/CME/documents/SM321-20-1-.pdf.

Shangri-La Group. "Our Heritage." Accessed June 13, 2022. https://www.shangri-la.com/group/our-story/our-heritage.

Shukla, Tarun. "Low Cost Carriers in India" In: Sven Gross and Michael Lück, eds. *The Low Cost Carrier Worldwide*. Abingdon, Oxon, UK: Routledge 2014. 139–154.

Simpfendorfer, Ben. *The Rise of the New East: Business Strategies for Success in a World of Increasing Complexity*. New York: Palgrave McMillan, 2014.

Singapore Redevelopment Authority. "The Marina Bay Story." Accessed July 2, 2022. https://www.ura.gov.sg/Corporate/Get-Involved/Shape-A-Distinctive-City/Explore-Our-City/Marina-Bay/The-Marina-Bay-Story.

Singapore Tourism Board. *Tourism Development in Singapore: A Report*. Singapore: Singapore Tourism Board, 1986.

Snaije, Olivia Snaije. "My Airport: Beirut International Airport; Nostalgia and a Dream of Unity." *Popula*. Accessed January 14, 2023. https://popula.com/2018/12/17/my-airport-beirut-international-airport/.

Sobocinska, Agnieszka. *Visiting the Neighbors: Australians in Asian*. Sydney: Newsouth, 2014.

Strand Film Company, Ltd. "Air Outpost." Newsreel. Accessed June 22, 2022. https://www.youtube.com/watch?v=253vJ6uYpq4.

Sudjic, Deyan. *The Edifice Complex: How the Rich and Powerful Shape the World*. New York: The Penguin Press, 2005.

Sumers, Brian. "Interview: Jetstar CEO on Running the World's Most Unique Low-Cost Carrier." *Skift*. Accessed January 17, 2023. https://skift.com/2016/09/21/ceo-interview-running-jetstar-group-the-worlds-most-unique-low-cost-carrier/.

Tacke, Heinfried and Sabina Marreiros, eds. *Dubai: Architecture & Design*. Susan Kirkbright, trans. Cologne, Germany: Daab, 2006.

Taj Mahal Hotels. *The Centenary: Taj, 100 Years of Glory*. Bombay, 1993.

Taliesin Associated Architects. *Minoo and Kish Islands: Potentially Important Resorts for Modern Iran, Stage I*. Scottsdale, AZ: The Frank Lloyd Wright Foundation, November 30, 1967.

Taliesin Associated Architects. *The Islands of Minoo and Kish: Tourist Feasibility Analysis and Report for the Ministry of the Interior, Stage 2*. Volume 2. Scottsdale, AZ: The Frank Lloyd Wright Foundation, August 1968.

Taumoepeau, Sesimi. "Suitability of the Low-Cost Airline Model in the South Pacific Region." In: David Timothy Duval ed. *Air Transport in the Asia Pacific*. Farnham, UK: Ashgate, 2014. 93–112.

Taumoepeau, Semisi. "Low Cost Carriers in the Asia Pacific." In: Sven Gross and Michael Lück, eds. *The Low Cost Carrier Worldwide*. Abingdon, Oxon, UK: Routledge, 2014. 113–138.

Tang, Chuanzhong. "Exploring the potential of hub airports and airlines to convert stopover passengers into stayover visitors: Evidence from Singapore." PhD dissertation. Department of Tourism, Sport and Hotel Management Griffith Business Group. Griffith University. Accessed January 17, 2023. https://research-repository.griffith.edu.au/bitstream/handle/10072/366161/Tang_2015_02Thesis.pdf?sequence=1.

Tata, JDR. *Letters* Arvind Mambro, ed. New Delhi: Rupa & Company, 2004.

Turnbull, Constance. *History of Modern Singapore, 1819–2005*. Singapore: National University of Singapore, 2020.

UNESCO. *Iran: Preservation et Mise en Valeur du Patrimoine Monumental de L'Iran en Liaison avec le Developpement du Tourisme dans ce Pays*. Paris: UNESCO, June 1967.

Vishwajeet, Shelley. The *IndiGo Story: Inside the Upstart that Redefined Indian Aviation*. Rupa Publications India Pvt. Ltd, 2018.

Volgyi, Katalin. "A Successful Model of State Capitalism: Singapore." In: Miklos Szanyi, ed. *Seeking the Best Master: State Ownership in the Varieties of Capitalism*. Central European Press, 2019. 276–296.

Vora, Neha. *Impossible Citizens: Dubai's Indian Diaspora*. Durham, North Carolina: Duke University Press, 2013.

Vowles, Timothy andDaniel Mertens. "Gateway Airports and International and Regional Connectivity of Air Transport in the Asia Pacific," In: David Timothy Duval ed. *Air Transport in the Asia Pacific*. Farnham, UK: Ashgate, 2014.

Waldron, Greg. "Strategy: Qantas to revisit 'Project Sunrise' at end 2021: Alan Joyce." Accessed January 29, 2022. https://www.flightglobal.com/strategy/qantas-to-revisit-project-sunrise-at-end -2021-alan-joyce/141936.article.

Walker, David. *Anxious Nation: Australia and the Rise of Asia, 1950–1939*. Brisbane: University of Queensland Press, 1999.

Ward, Evan R. "Before Dubai: The Frank Lloyd Wright Foundation and Iranian Tourism Development, 1967–1969," *Journal of Tourism History* 11 (2019). Accessed January 14, 2023. https://www.tandfon line.com/doi/full/10.1080/1755182X.2019.1567830?src=recsys.

Wesley, Michael. *There Goes the Neighborhood: Australia and the Rise of Asia*. Sydney: New South, 2012.

Wharton, Annabel. *Building the Cold War: Hilton International Hotels and Modern Architecture*. Chicago: University of Chicago Press, 2001.

Wilson, Graeme. *Emirates: The Airline of the Future*. London: Prima Media, 2007.

Wilson, Graeme. *Flight into the Future: Seventy Years of Civil Aviation in Dubai*. London: Media Prima, 2007.

Wilson, Graeme. *Fly Buy Dubai: The Remarkable 25 Year Journey of Dubai Duty Free*. London: Media Prima, 2008.

Wilson, Graeme. *Rashid's Legacy: The Genesis of the Maktoum Family and the History of Dubai*. London: Media Prima, 2006.

World Tourism Organization. *Statistical Report on the Period, 1967–1976*. Madrid: World Tourism Organization, 1977.

Wong, John D. *Hong Kong Takes Flight: Commercial Aviation and the Making of a Global Hub, 1930s–1998*. Cambridge, Massachusetts: Harvard University Press, 2022.

Wood, Robert E. "Tourism and Underdevelopment in Southeast Asia." *Journal of Contemporary Asia* 9 (1979).

Wright, Nadiad H. *The Armenians of Singapore: A Short History* (Penang, Malaysia: Entrepot Publishing, 2015.

Yeo Cheow Tong. "Speech by Mr. Yeo Cheow Tong at the Annual Airport Reception on 13 January 2001." Accessed June 15, 2021. https://www.mot.gov.sg/news-centre/news/Detail/ Speech%20by%20Mr%20Yeo%20Cheow%20Tong%20at%20the%20Annual%20Airport%20Recep tion%20on%2013%20January%202001/.

Young, Gavin. *Beyond Lion Rock: The Story of Cathay Pacific Airways*. London: Hutchinson, 1988.

Young, Kenneth T. "Asia and America at the Crossroads." *The Annals of the American Academy*.

Yu Chun-ho. "Information Note: Policy Measures to Enhance Airport Competitiveness in Selected Places." Legislative Council Secretariat Research Office. Accessed January 19, 2022. https://www. legco.gov.hk/research-publications/english/1718in01-policy-measures-to-enhance-airport- competitiveness-in-selected-places-20171027-e.pdf.

Index

https://doi.org/10.1515/9783111326641-009

Bei Fragen zur Produktsicherheit wenden Sie sich bitte an:
If you have any questions regarding product safety,
please contact:

Walter de Gruyter GmbH
Genthiner Straße 13
10785 Berlin
productsafety@degruyterbrill.com